W0235425

Knowing the Structure of Nature

*Also by Stathis Psillos*

THE ROUTLEDGE COMPANION TO PHILOSOPHY OF SCIENCE (*co-edited with Martin Curd, 2008*)
*This book was one of Choice Outstanding Academic Titles, for 2008.*

PHILOSOPHY OF SCIENCE A–Z (2007)

CAUSATION AND EXPLANATION (2002)
*This book was awarded the British Society for the Philosophy of Science Past Presidents' Prize for best textbook in the philosophy of science, for 2004.*

SCIENTIFIC REALISM: HOW SCIENCE TRACKS TRUTH (1999)

# Knowing the Structure of Nature

## Essays on Realism and Explanation

Stathis Psillos
*University of Athens*

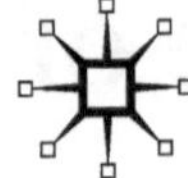



First published 2009 by
PALGRAVE MACMILLAN

Palgrave Macmillan in the UK is an imprint of Macmillan Publishers Limited, registered in England, company number 785998, of Houndmills, Basingstoke, Hampshire RG21 6XS.

Palgrave Macmillan in the US is a division of St Martin's Press LLC, 175 Fifth Avenue, New York, NY 10010.

Palgrave Macmillan is the global academic imprint of the above companies and has companies and representatives throughout the world.



ISBN-13: 978–0–230–00711–6 hardback

A catalogue record for this book is available from the British Library.

A catalog record for this book is available from the Library of Congress.

10 9 8 7 6 5 4 3 2 1
18 17 16 15 14 13 12 11 10 09

Transferred to Digital Printing in 2013

*for my younger daughter, Artemis*
*another Olympian goddess of my life*

# Contents

# Preface

This book was conceived of as a collection of essays – papers that have been composed between 2000 and 2005 (and published as late as in 2008) in the shadow of my book *Scientific Realism: How Science Tracks Truth*, which was published in 1999. Though it is still a collection of essays (all but one chapter make their second appearance here with relatively minor stylistic changes and corrections), it has acquired a lot more unity and cohesion than I first expected. This is partly due to the fact that the essays of this volume are mostly about scientific realism and inference to the best explanation.

A number of people that I can identify and many others personally unknown to me who attended talks and seminars I have given over the years have helped me with (oral and written) comments, criticism and advice. Without them, I would have made minimal progress with the ideas and problems presented in this book. Still, it is left to the readers to judge whether I have managed to capitalise on all these people's generous help and have indeed made some progress. If I have not, *mea culpa*.

Here is a (hopefully near complete) list of people that I want to thank in person. D. A. Anapolitanos, Theodore Arabatzis, Jody Azzouni, Alexander Bird, Otavio Bueno, Craig Callender, Nancy Cartwright, Anjan Chakravartty, Hasok Chang, Paul Churchland, Peter Clark, Steve Clarke, Chris Daly, Bill Demopoulos, Gerald Doppelt, Mauro Dorato, Igor Douven, Alice Drewery, John Earman, Brian Ellis, Jan Faye, Steven French, Maria Carla Galavotti, Michel Ghins, Ron Giere, Olav Gjelsvik, Barry Gower, Bengt Hansson, Stephan Hartmann, Robin Hendry, Carl Hoefer, Leon Horsten, Mikael Karlsson, Jeff Ketland, Bob Kowalski, James Ladyman, Elaine Landry, Oystein Linnebo, Peter Lipton, Peter Machamer, Uskali Maki, D. H. Mellor, Ilkka Niiniluoto, Robert Nola, John Norton, David Papineau, Nils Roll-Hansen, Nils-Eric Sahlin, Howard Sankey, Christoff Schmidt-Petri, Matti Sintonen, Brian Skyrms, Wolfgang Spohn, David Spurrett, P. Kyle Stanford, Mauricio Suarez, Patrick Suppes, Paul Teller, Thomas Uebel, Bas van Fraassen, Janneke van Lith, Stelios Virvidakis, Ioannis Votsis, Dag Westerstahl, Asa Wikforss and John Worrall. I owe all of you at least a pint of real ale!

The same goes for the anonymous readers for journals and books in which the papers of this volume have appeared – if only I knew who they were.

Here is a list of audiences in seminars that I want to thank – something akin to the anonymous philosopher.

*Departments of Philosophy*: Bogazici University; Erasmus University of Rotterdam; University of California, San Diego; University of Ghent; University of Gothenburg; University of Helsinki; University of Lund; University of Oslo; University of Pittsburgh (HPS) and the University of Stockholm.

*Conferences and Workshops*: the annual conference of the *British Society for the Philosophy of Science* (University of Sheffield, July 2000); the 17th Biennial Meeting of the Philosophy of Science Association (Vancouver, November 2000); the NORDPLUS Intensive Programme on Inference to the Best Explanation (Iceland, September 2001); the Workshop 'Induction and Deduction in the Sciences' (Vienna, July 2002); the annual conference of the *British Society for the Philosophy of Science* (University of Glasgow, July 2002); the Workshop in honour of Nancy Cartwright (University of Konstanz, December 2002); the Vienna Circle Institute Ramsey Conference (Vienna, November 2003); the Belgian Society for Logic and Philosophy of Science (Brussels, February 2004); the Ratio Conference (University of Reading, May 2004); the 19th Biennial Meeting of the Philosophy of Science Association (Austin Texas, November 2004); the annual conference of the *British Society for the Philosophy of Science* (University of Bristol, July 2007).

The idea for a book of this sort belonged to Dan Bunyard, who used to be the assistant editor in Philosophy and History for Palgrave/Macmillan and now has moved on to a different press. I want to thank him wholeheartedly for all his patience, encouragement and support. My thanks should be extended to Priyanka Pathak who took over from Dan at Palgrave/Macmillan. Pri (and Melanie Blair, Senior Editorial Assistant at Palgrave) showed great care for this project, but most of all, extreme patience with my failures to meet one deadline after the other. Many thanks to my student Aspassia Kanellou for her generous help with compiling the references.

Part of the research that goes into this volume was funded by the Research Programme 'Kapodistrias' of the University of Athens.

My family deserves my gratitude, for their unfailing support and love. There is more to life than books, as my older daughter Demetra says. Athena, Demetra and the newcomer Artemis prove to me that this more is worth more than all books taken together – and more!

Here is an acknowledgement of the provenance of the chapters of this book.

Chapter 1: First published as 'The Present State of the Scientific Realism Debate' *The British Journal for the Philosophy of Science* 51 (Special Supplement), 2000, pp.705–28.

Chapter 2: First published as 'Scientific Realism and Metaphysics', *Ratio* 18, 2005, 385–404. With the kind permission of Wiley-Blackwell.

Chapter 3: First published as 'Thinking About the Ultimate Argument for Realism', in Colin Cheyne & John Worrall (eds) *Rationality & Reality*, 2006, Springer, pp. 133–56. With kind permission of Springer Science and Business Media.

Chapter 4 appears here for the first time.

Chapter 5: First published as 'Tracking the Real: Through Thick and Thin', *The British Journal for the Philosophy of Science* 55, 2004, pp. 393–409.

Chapter 6: 'Cartwright's Realist Toil: From Entities to Capacities', in Stephan Hartman, Carl Hoefer & Luc Bovens (eds) *Cartwright's Philosophy of Science*, 2008, Routledge, pp. 167–94. With the kind permission of Taylor & Francis Group.

Chapter 7: First published as 'Is Structural Realism Possible?' *Philosophy of Science* 68, 2001, pp. S13–24 © 2001 by the Philosophy of Science Association.

Chapter 8: First published as '*The* Structure, the *Whole* Structure and Nothing *but* the Structure?', *Philosophy of Science* 73, 2006, pp. 560–70. © 2006 by the Philosophy of Science Association.

Chapter 9: First published as 'Ramsey's Ramsey-Sentences', in Maria Carla Galavotti (ed.) *Cambridge and Vienna: Frank P Ramsey and the Vienna Circle*, 2006, Springer, pp. 67–90. With kind permission of Springer Science and Business Media.

Chapter 10: First published as 'Simply the Best: A Case for Abduction' in A. C. Kakas & F. Sadri (eds) *Computational Logic: From Logic Programming into the Future*, LNAI 2408, Berlin–Heidelberg: Springer–Verlag, 2002, pp. 605–25. With kind permission of Springer Science and Business Media.

Chapter 11: First published as 'Inference to the Best Explanation and Bayesianism' in F. Stadler (ed.) *Induction and Deduction in the Sciences*, Kluwer, 2004, pp. 83–91. With kind permission of Springer Science and Business Media.

# Introduction

The title of this book is inspired by a passage of the *Critique of Pure Reason* (A278/B334) in which Kant talks about knowledge of the secrets (the inner recesses) of nature. Knowledge (at least of the sort licensed by the theoretical reason), Kant thought, is always knowledge of the appearances, which are constituted by the forms of pure intuition (space and time) and the schematised categories of the understanding. The things-as-they-are-in-themselves will remain forever hidden to the human intellect, bound as it is by sensible intuition. Being minded as we are forecloses the possibility of getting epistemic access to the noumena. This kind of dichotomy between something inherently unknowable and something knowable might well be based on epistemic modesty – which has been known as Kantian Humility.[1] This kind of epistemic dichotomy requires a certain epistemic disconnection between whatever is necessary for getting to know nature (the appearances), of which Kant said that 'no one knows how far this knowledge may in time extend' and what would be necessary for knowing the transcendental objects which may well be the ground (or the cause, as Kant occasionally says) of the appearances – those of which Kant said they would remain unknowable even if 'the whole of nature were revealed to us'. It also requires an argument for positing, nonetheless, something that remains forever beyond our ken. The ways Kant tried to meet these two requirements are too well known to bear recounting.

What does need recounting is other attempts to draw a sharp epistemic dichotomy between those aspects of nature that are knowable and those that will remain secret – attempts to circumscribe realms of being that are cognitively impenetrable to us, to set limits to our knowledge of nature.

The currently most popular epistemically dichotomous position is structural realism. The epistemic dichotomy structural realism introduces is between knowing the structure of nature and knowing whatever is left to fill out the structure. Knowing the structure of nature is deemed possible, but that's where the epistemic optimism ends: there is something else in the world (another realm of being) whose knowledge is not deemed possible. To call this extra stuff non-structure might sound trivial, but it is helpful nonetheless since there is a good deal of vagueness and haziness in calling it 'nature' (structure vs nature) or 'content' (structure vs content). The structuralist dichotomy might well have some Kantian origin and inspiration (it did have for Henri Poincaré and Bertrand Russell, for instance). The non-structural unknowable part of nature may well be much like the Kantian noumenon. There is no reason, however, for structuralism to endorse all the heavy-weight Kantian metaphysics, let alone all the stuff about intuition and its pure forms.

Epistemic dichotomies are introduced by other philosophical positions as well. In all cases, a principled line is drawn between whatever there is warranted epistemic access to and whatever is merely speculative or conjectural or simply unwarranted. Entity realism, of the sort advocated by Nancy Cartwright, is one such position. And so is the call for thick epistemic access to the real, advanced by Jodi Azzouni.

The main aim of this book is to defend the epistemic optimism associated with scientific realism. It does that without imposing (actually, with denying) any epistemic dichotomies; without implying or requiring (actually, with denying) that there are principled limits in knowing the secrets of nature. Well, there is no a priori reason to think that the human intellect will be able to know all the secrets of nature. But then again there is no good reason (a priori or a posteriori) to think that there is a principled epistemic division between what can be known and what cannot: the forever secrets of nature (if there are such) do not fall nicely within a conceptual category (the unknown X: the noumena; the non-structure; the intrinsic properties; the auxiliary properties; whatever-there-is-only-thin-epistemic-access-to; whatever-there-is-only-theory-mediated-access-to; and the like). Or so I will argue, anyway, in the various chapters of this book. Hence, the title of the book has a pinch of irony. Knowing the *structure* of nature is indeed possible, but beware of undesired connotations: it's not as if there is something left over that is inherently unknowable; some in principle (and for principled reasons) secret of nature.

The book is divided into three parts. The first (and longest) continues the articulation and defence (with corrections and revisions, where necessary) of scientific realism, as this was first developed in my *Scientific Realism: How Science Tracks Truth* (1999). The second part is about structural realism: it criticises both the epistemic and the ontic version of structuralism and offers a reconstruction of Frank Ramsey's account of scientific theories, distancing his approach from structuralism. The third (and shortest) part is on inference to the best explanation (IBE). It sketches a framework within which the issue of the defence of the reliability of IBE can be raised and answered – a framework based on John Pollock's innovative work in epistemology. It also makes some rather militant points against the current orthodoxy in matters of methodology, namely, subjective Bayesianism.

The first part begins with a detailed survey of the scientific realism debate circa 2000. Echoing Imre Lakatos, one could dub the 1990s the 'Renaissance of scientific realism in the philosophy of science'. Chapter 1 discusses the main argumentative strategies in defence of scientific realism, the several strands within realism and the key issues involved in the formulation of scientific realism as a philosophical position. Scientific realism is based on a double claim – a declaration of independence (the world science aims to know is mind-independent) and a call for epistemic optimism (this mind-independent world is knowable). This is humility and audacity side by side. To many, this combination is bankrupt – hence, either humility or audacity has to be (occasionally reluctantly) sacrificed. Humility (the independence of the world from the knowing subject) is retained but is accompanied by ignorance (by virtue of being independent of the knowing subject, this world – or some aspects of it – are not knowable). Alternatively, audacity (the world is knowable) is retained, but accompanied by arrogance (by virtue of being knowable, this world is not mind-independent). Verificationist realists go for the second option, sceptical realists go for the first. Part of the argument of Chapter 1 is that scientific realists are not forced to go either for verificationism or for scepticism. Epistemic optimism need not go against the declaration of independence. An important issue, however, is how exactly the so-called mind-independence of the world should be understood. In Chapters 1 and (part of chapter) 2, it is suggested that the best way to understand the claim of mind-independence of the world is via commitment to a non-epistemic conception of truth.

One important issue in the scientific realism debate is how rich the metaphysics of scientific realism should be. This is, normally, a family quarrel – fought out among scientific realists who take it for granted that a basic realist commitment is to an independently and objectively existing reality. But what about the metaphysical constitution of this reality? Is it a reality that comprises universals and powers and *de re* necessities and relations of necessitation and essential properties and all that stuff that is in vogue lately and constitutes the building blocks of a non-Humean conception of the deep structure of the world? I think there are independent reasons to resist this non-Humean conception of reality. Some of them are explained in Chapter 6, where I take issue with Cartwright's hyper-realism about capacities and powers. Even if we disregard these reasons, a point stressed in Chapter 2 is that commitment to scientific realism does not imply *deep* metaphysical commitments.

In particular, it does not imply commitment to physicalism or to a non-Humean metaphysics. One can take a rich view of reality as comprising all facts. Alternatively, one may take a more austere view of reality, according to which only an elite class of facts (the fundamental facts) is real (*really* real, if you like). How exactly these elite facts are circumscribed is a separate issue. Scientific realism should not be confused with fundamentalism. Actually, it is a confusion like this that fosters the thought that scientific realism is a richer metaphysical view than it is (ought to be). My diagnosis is that realism should adopt a factualist and not a fundamentalist view of reality. Then, one can be realist about a number of domains (or subject-matters) without *ipso facto* taking a stance on independent metaphysical issues. The issue of whether some entities are basic, or derivative, or irreducible, or *sui generis* is *separate* and needs to be addressed separately. The stances one might take will stem from *other* metaphysical commitments one might have, for example, a commitment to physicalism, or naturalism, or pluralism.

What warrants the epistemic optimism associated with scientific realism? The main argument has been the so-called no-miracles argument. Briefly put, it is an argument that mobilises the successes of scientific theories (especially their successful novel predictions) in order to suggest that the best explanation of these theory-driven successes is that the theories that fuelled them were approximately true – at least in those respects that were implicated in the generation of the successes. There are a number of niceties in the construction of this argument, but in the way I have formulated it in Psillos (1999), it aims to defend the reliability of scientific methodology in producing approximately true theories and hypotheses. Following more concrete types of explanatory reasoning that occur all the time in science, it suggests that it is reasonable to accept certain theories as approximately true, at least in the respects relevant to their theory-led predictions. These successful instances of explanatory reasoning in science provide the *basis* for a grand abductive argument. The no-miracles argument, however, is not just a generalisation over the scientists' abductive inferences. Although itself an instance of the method that scientists employ, it aims at a much broader target: to defend the thesis that IBE is reliable.

The main objection to this argument hits directly its core, namely, that it is an instance of inference to the best explanation aiming to vindicate inference to the best explanation; hence, it is viciously circular and therefore epistemically impotent. As I argued in Psillos (1999), there is a plausible distinction between premise-circularity and rule-circularity (a premise-circular argument employs its conclusion as one of its premises; a rule-circular argument conforms to the *rule* which is vindicated in its conclusion). What is more, (a) the abductive defence of realism is rule-circular, (b) rule-circularity is *not* vicious and (c) rule-circularity is involved in the defence of all basic and fundamental rules of inference.

The no-miracles argument is revisited in Chapter 3. There the emphasis is on attempts to dismiss it as either deductively invalid or inductively invalid argument. Being committed to deductivism, Alan Musgrave has argued that the no-miracles argument is best understood as a deductive enthymeme. Being committed to inductivism, Colin Howson has argued that the no-miracles argument, as usually understood, falls foul of the base-rate fallacy and has proposed that it is best understood as a Bayesian argument with an explicit reliance on subjective prior probabilities. A detailed examination of both types of reaction to the no-miracles argument suggests that – seen as an instance of IBE – it survives criticism. In any case, the right way to proceed is to look at case histories, namely, concrete cases of explanatory and predictive successes in science. Then, there are strong best-explanation-based reasons to be realists about several theories.

Instrumentalism, one of the chief rivals of scientific realism, has been a broad church. In its strictest sense, as defended by Philipp Frank (1932), it is a form of non-cognitivism. He took it that the aim of science is prediction and that theories are merely tools for this aim – in particular, symbolic tools that do not (aim to) represent the real world. Theories, for Frank, are neither true nor false; they have only instrumental value, which is cashed in terms of predictions of future observations on the basis of present ones. In this form, instrumentalism is now discredited. The realist no-miracles argument is to be credited for this feat.

There is a version of instrumentalism advocated by John Dewey, who I think coined the very term, who took it to be the view licensed by pragmatism. For him a deep philosophical problem was the relation between the 'conceptual objects' of science and the things of ordinary experience. Dewey seems to have favoured some kind of contextualism: reality is not an absolute category firmly attributed to some entities and firmly denied to some other. As he put it: 'That the table *as* a perceived table is an object of knowledge in one context as truly as the physical atoms, molecules, etc. are in another situational context and with reference to another *problem* is a position I have given considerable space to developing' (Dewey 1939, 537). The context and the problem are determined, at least partly, by the things one *does* with an entity and by the role an entity plays within a system – one cannot put books on swarm of molecules, as Dewey says. One may well question the motivation for this view (since it is not clear, to say the least, how contexts are separated). But here, in any case, is a variant of instrumentalism that does not reduce theories to calculating devices; nor does it deny that (in a sense, at least) explanatory posits are real.

I take it that P. Kyle Stanford's (2006) neo-instrumentalist approach is a sophisticated version of contextualist instrumentalism. He draws a distinction between those entities to which there is an independent route of epistemic access (mediated by theories that cannot be subjected to serious doubt) and those entities to which all supposed epistemic access is mediated by high-level theories. He takes a realist stance on the former but he is an instrumentalist about the latter. High-level theories are taken to be useful conceptual tools for guiding action rather than maps of an unavailable to the senses reality. But his overall attitude is that the difference between realism and instrumentalism is local rather than global. It concerns the range of theories one is willing to take as true on the basis of the total evidence available.

This kind of attitude is motivated by what Stanford has called the 'new induction' on the history of science, according to which past historical evidence of transient underdetermination of theories by evidence makes it likely that current theories will be supplanted by hitherto unknown ones, which, nonetheless, are such that when they are formulated, they will be at least as well confirmed by the evidence as the current ones. The new induction is effective, if at all, only in tandem with the pessimistic induction. Hence, Stanford has taken extra pains to show that the arguments against the pessimistic induction advanced by Philip Kitcher and myself are failures. Chapter 4 of the book analyses and criticises Stanford's arguments and casts doubt on his neo-instrumentalist outlook. As in many other cases of instrumentalism, Stanford's is motivated by an epistemic dichotomy between the life-world and the image of the world as it is depicted by science, and especially by high-level theories. Stanford recasts this dichotomy in interesting ways (allowing the life-world to have a much richer content than strict empiricists would, including a host of unobservable denizen). But, ultimately, he too rests his case against full-blown realism on dodgy double standards in confirmation.

The scientific realist turn in the 1960s was spearheaded by Wilfrid Sellars's critique of the myth of the given and his adjoining critique of the layer-cake view of scientific explanation. This view presupposed that the role played by the theoretical entities was to explain the inductively established empirical laws that were formulated by means of exclusively observational vocabulary. Sellars famously argued that the layer-cake view was grist to the instrumentalist mill precisely because, if you think of it, it renders the theoretical entities explanatorily dispensable: all the explanatory work is done by the empirical generalisations. The alternative, Sellars thought, was to establish the explanatory indispensability of theoretical entities and this is achieved by showing that theoretical entities explain directly individual and singular observable phenomena and *correct* the inductively established empirical generalisations. The observational framework does not have the resources to express strict laws that explain the observable phenomena.

This Sellarsian argument is revisited in Chapter 5. There it is contrasted to Azzouni's 'tracking requirement' according to which (roughly) the real should be epistemically tracked by thick epistemic processes, such as observation and instrument-based detection. This kind of thick epistemic access is supposed to be robust enough to enable tracking of the properties of objects, by virtue of which these objects become known. It is contrasted to thin epistemic access, which is access to objects and their properties via a *theory*, which is (holistically) confirmed and has the Quinean virtues (simplicity, familiarity, scope, fecundity and success under testing). Azzouni's thought is that 'thin access' is no access at all, since there is no robust connection between being real and being implied by a well-confirmed theory. A central point of Chapter 5 is that there are no self-authenticating observational (and instrument-based) epistemic processes. Since their epistemic authentication comes from theories, all these theories have to be themselves authenticated, that is *confirmed*. It follows that the very possibility of thick epistemic access to the real presupposes thin epistemic access to it.

The idea of thick epistemic access to the real resonates with the basic thesis of entity realism, namely, that it is epistemically safer to be realist about certain entities than to be about (high-level) scientific theories. Cartwright's version of entity realism is tied to causal explanation and, in particular, to the thought that causal explanation implies commitment to the entities that do the causing and hence the explaining. Chapter 6 discusses Cartwright's version of realism in detail. Cartwright aimed to drive a wedge between explanation and truth. Along the way she argued that law-statements (expressing fundamental laws of nature) are either true but non-explanatory or explanatory but false. She objected to inference to the best explanation and substituted for it what she called 'inference to the most likely cause'. It is argued, in Chapter 6, that by insisting on the motto of 'the most likely cause', Cartwright underplayed her argument for realism. She focused her attention on an *extrinsic* feature of causal inference, that is the demand of high probability, leaving behind the *intrinsic* qualities that causal explanation should have in order to provide the required understanding. Besides, though it is correct to claim that ontic commitment flows from causal explanation, she was wrong to tie this commitment solely to the entities that do the causal explaining. This move obscured the nature of causal explanation and its connection to laws.

As already noted, Cartwright has insisted on a non-Humean view of the world, being taken to be populated by capacities and active natures. In fact, her argument for capacities is taken to mirror Sellars's powerful argument for the indispensable explanatory role of unobservable entities. The later section of Chapter 6 scrutinise Cartwright's non-Humean conception of the world and cast doubt on it and on its defence.

The second part of the book is devoted to structuralism. There have been two distinctive strands in epistemic structuralism. The first has been motivated by historical considerations that have to do with theory-change. It has found its classic expression in the writings of Poincaré and Pierre Duhem. Poincaré took the basic axioms of geometry and mechanics to be (ultimately, freely chosen) conventions. He nonetheless thought that scientific hypotheses proper, even high-level ones such as Maxwell's laws, were empirical. But theories tend to come and go. There seems to be radical discontinuity in the transition from old theories to new ones: some central and basic scientific hypotheses and law-statements are abandoned. Others, prima facie incompatible with the latter are adopted. Poincaré noted that a closer look at theory-change in science reveals a pattern of substantial continuity at the level of the mathematical equations that represent empirical as well as theoretical *relations*. For him, these retained mathematical equations – together with the carried-over empirical content – fully capture the objective content of scientific theories. By and large, he thought, the objective theoretical content of scientific theories is structural: if successful, a theory represents correctly the *structure* of the world. In the end, the structure of the world is revealed by structurally convergent scientific theories. Duhem articulated further this idea by drawing a distinction between the explanatory part and the representative part of a scientific theory. The former contains hypotheses as to the underlying causes of the phenomena, whereas the latter comprises the empirical laws as well as the mathematical formalism which is used to represent, systematise and correlate these laws. Duhem's claim was that in theory-change, the explanatory part of a theory gets abandoned while its representative (structural) part gets absorbed by the successor theory. This structural convergence is meant to be conducive to the aim of science, which for Duhem was the natural classification of the phenomena – where a classification (that is, the organisation of the phenomena within a mathematical structure) is *natural* if the relations it establishes among the experimental phenomena captures real relations among things.

The second strand in epistemic structuralism is impervious to history and is motivated by the thought that empiricism can be reconciled with the possibility of knowledge that goes beyond sensory evidence and experience. This view has found its defining moment in Russell's structuralist account of the knowledge of the world. According to this, only the formal structure, that is, the totality of logico-mathematical properties, of the world can be known, while the intrinsic properties of the unobservable causes of the perceived phenomena are inherently unknown. This logico-mathematical structure, Russell claimed, can legitimately be inferred from the structure of the perceived phenomena. This kind of view was developed further by Grover Maxwell, who linked it with the Ramsey-sentence representation of scientific theories and updated it by dropping the requirement of inferential knowledge of the structure of the unobservable world. Maxwell defended structural realism as a form of representative realism, which suggests that (i) scientific theories imply commitment to unobservable entities but (ii) all knowledge of these unobservables entities is structural, that is, knowledge of their higher-order properties and not of their first-order (or intrinsic) properties.

Despite their interesting differences (for instance, concerning the notion of structure), both structuralist strands have been fused to form the backbone of the best current version of epistemic structuralism, as this has been expressed in the writings of John Worrall and Elie Zahar. On this view, the empirical content of a theory is captured by its Ramsey-sentence while the excess content that the Ramsey-sentence has over its observational consequences is purely *structural*. Given that the Ramsey-sentence captures the logico-mathematical form of the original unramseyfied theory, the epistemic structuralist thought is that, if true, the Ramsey-sentence also captures the structure of reality.

This sketchy outline of a rich research programme is filled out in Chapters 7–9. Collectively, these chapters aim to rebut the epistemic structuralist dichotomy: there is no cleavage between knowing the structure of nature and knowing something beyond or behind it (the non-structure of nature, as it were). The structuralist attempt to raise a principled epistemic barrier in our knowledge of the world, limited to whatever is captured by the Ramsey-sentences of our best scientific theories, is ill-motivated and wrong-headed. The key objection to epistemic structuralism goes back to an important point made by the Cambridge mathematician Max Newman in reply to Russell's structuralism, but is affective against all forms of epistemic structuralism: the structuralist claim that *only* structure can be known is either trivial or false. The importance of the Newman problem is discussed in detail in Chapters 7 and 9.

The failures of epistemic structuralism have led many structuralists to make an ontic turn. They take structuralism to be an ontic position which asserts that structure is all there is. This is a slogan, of course, and can be (and has been) fleshed out in many ways. Whichever way one goes, the bottom line of ontic structuralism is a denial of the epistemic dichotomy associated with epistemic structuralism. If structure is all there is, there is nothing leftover to remain in principle unknowable, always hidden from us – an eternal secret of nature. This is certainly a welcome conclusion from the perspective of the present book. But to say that *structure* is all there is may well flirt with vacuity because if there is no distinction to be drawn between structure and something else, there is no sense in which all there is is *structure* – structure as opposed to what? As van Fraassen (2006, 18) has aptly put it: 'it seems that, once adopted, [ontic structuralism] should not be called structuralism at all! For if there is no non-structure, there is no structure either.'

A way out of this almost paradoxical situation is to think of ontic structuralism as a call to do away with objects as individuals that exist and are individuated independently of relational structures – objects that are not themselves structures. This is not far from the Kantian view – at least as Ernst Cassirer has read it: objects consist entirely of relationships; except that ontic structuralists lose the Kantian noumena. A central motivation for ontic structuralism comes from problems in the interpretation of quantum mechanics and the metaphysics (and the physics) of individuality. This is an area I do not have much to say about – mostly due to ignorance. The bulk of Chapter 8 aims to show that ontic structuralism lacks the resources to accommodate causation within its structuralist slogan. It is also argued that some substantive non-structural assumption needs to be in place before we can talk about *the* structure of anything. Drawing on some relevant issues concerning mathematical structuralism, it is claimed that (a) structures *need* objects and (b) scientific structuralism should focus on *in re* structures. These – *aka* relational systems – are concrete spatio-temporal entities: structures are no longer freestanding entities. If ontic priority is given to *in re* structures, there is more to the world than structure.

As noted already, the epistemic structuralist movement has found in Ramsey-sentences a way to draw an epistemically relevant distinction between structure and whatever 'fills in' the structure. Chapter 9 offers a systematic analysis of Ramsey's seminal essay *Theories*, that was published posthumously in 1931 and received a lot of attention in the 1950s and beyond. Ramsey's own views of Ramsey-sentences have both historical and philosophical significance. There have been a number of different ways to read Ramsey's views and to reconstruct Ramsey's project – one of them being the epistemic structuralist way. A key claim of Chapter 9 is that, for good reasons, Ramsey did not see his Ramsey-sentences as part of some sort of structuralist programme.

My own interpretation of Ramsey's Ramsey-sentences, which I think Ramsey might have found congenial, takes theories to be growing existential statements. In adopting a theory we commit ourselves to the existence of things that make it true and, in particular, to the existence of unobservable entities that cause or explain the observable phenomena. We also take it to be the case that the world (or the domain of the theory) has a concrete structure (its natural joints). We do that for many reasons, but a prominent among them is that we want to render substantive our theory's claim to truth (and to leave open the possibility of its falsity). We don't want our theory to be true just in case it is empirically adequate. The natural (causal-nomological) structure of the world should act as an *external constraint* on the truth or falsity of our theory. This bold attitude is balanced by some sort of humility. We don't foreclose the possibility that what the things we posited are might not be found out. Some things must exist if our theory is to be true and these things must possess a natural structure if this truth is substantive. But humility advises us that there are many ways in which these commitments can be spelt out. Whatever else we do, we do not draw a sharp and principled distinction between what can be known and what cannot. We are not lured into thinking that only the structure of the unobservable world can be known, or that only the structural properties of the entities we posited are knowable or that we are cognitively shut off from their intrinsic properties. Humility need not lead us to accepting dogmatic epistemic dichotomies, where (there is good reason to think that) only continua exist. This sketchy image is what I call Ramseyan humility (which is independent and different from David Lewis's conception of it) and among other things it is meant to offer an olive branch to epistemic structuralists.

The third part of the book is focused on IBE. IBE is centrally involved in the defence of realism. It is also centrally involved in the formation and justification of beliefs in unobservable entities. But more needs to be said about IBE, especially in light of the fact that its defence should not be tied too closely to an externalist perspective in epistemology. IBE does need articulation and defence and the latter should be available even within an internalist perspective on justification, where justification and warrant are tied to the presence (or absence) of defeaters.

When it comes to ampliative reasoning in general, two are the issues we are faced with. The first is the issue of justification: given that ampliative reasoning is not necessarily truth-preserving, how can it be justified? The second is the issue of description: what is the structure of ampliative reasoning? The first is *Hume*'s problem; the second is *Peirce*'s problem. These two issues are distinct but a good philosophical theory of ampliative reasoning should aim to address both. In particular, a good philosophical theory should aim to show how some epistemic warrant is conferred on the outcomes of ampliative methods.

Chapter 10 shows that abduction, suitably understood as inference to the best explanation, offers the best description of scientific method. It is argued that both enumerative induction and the method of hypothesis (*aka* the hypothetico-deductive method) are limiting cases of IBE. It is also argued that there are ways to show that IBE is warrant-conferring. This is done by setting the whole issue of defending IBE within Pollock's (1986; 1987) framework of defeasible reasoning. By thinking of justification in terms of the absence of *defeaters*, this framework makes possible the investigation of the conditions under which defeasible reasoning can issue in warranted beliefs.

Chapter 11 focuses on the attempts to assimilate IBE within the confines of Bayesanism. Using somewhat polemic tone, I resist the temptation to cast IBE within Bayesianism. There are many reasons for this resistance, but a chief one among them is that seen within Bayesianism, IBE loses much of its excitement: it amounts to a *permission* to use explanatory considerations in fixing an initial distribution of prior probabilities.

I do not think that the type of inference broadly called 'defeasible', or 'ampliative', or 'non-monotonic' admits of a neat and simple abstract-logical form. Nor does it admit of an analysis suitable for deductive logic. Its details are very messy to admit of a neat abstract characterisation, but its strength lies precisely in those details. Unlike deductive reasoning and Bayesian updating (both of which are non-ampliative), defeasible reasoning has the prima facie advantage to take us to beliefs with new content. It does not start with the assumption that reasoning (and inference) is about updating an already existing (learned at our mother's knee?) belief corpus in light of new information (evidence). It plays an active role in creating this belief-corpus. Perhaps, a deductivist or a Bayesian could mimic defeasible or ampliative inference. As explained in Chapter 3, one could treat all inferences as deductive enthymemes or try to incorporate rich-in-content hypotheses in the belief corpus over which the prior probability distribution is fixed. But I see no motivation for this other than the (imperialist) claim that all reasoning should comply with a fixed and topic-neutral set of rules. Perhaps, one could argue that the so-called defeasible reasoning is an illusion, or a deviation from pre-established norms, or that it lacks justification and so on. But I see no motivation for this other than the question-begging claim that there is a logical tribunal to which all types of reasoning should appeal before they are real or justified or warranted. This patent-office view would surely invite the question of how the already patented forms of correct reasoning have acquired their credentials.

Presently (see Psillos 2007a), I think we should take a more contextualist view of IBE and approach the issue of its justification from another route. We should examine specific cases of defeasible reasoning in which explanatory considerations play a key role. These cases reveal that an IBE-type of reasoning has a fine structure that is shaped, by and large, by the context. Explanations are, by and large, detailed stories. The background beliefs rank the competitors. Other background assumptions determine the part of the logical space that we look for competitors. The relevant virtues or epistemic values are fixed and so on. Given this rich context, one can conclude, for instance, that the double-helix model is the best explanation of the relevant evidence, or that the recession of the distant stars is the best explanation of the red-shift. What *is* important in cases such as these is that the *context* can (a) show how explanatory considerations guide inference; (b) settle what is the relevant explanatory relation; (c) determine (in most typical cases) the ranking of the rival explanations; (d) settle what assumptions must be in place for the best explanation to be acceptable; (e) settle what to watch out for (or search) before the best explanation is accepted (e.g., the absence of certain defeaters). These contextual factors can link nicely loveliness and likeliness (to use the Late Peter Lipton's apt expressions), because they do not try to forge an abstract connection between them; rather the connection stands or falls together with the richness and specificity of the relevant information available.

It is only after we are clear on the rich fine structure of specific inferences that we can start extracting (or abstracting) a common inferential pattern of defeasible reasoning based on explanatory considerations. The key general idea is that IBE rests on an explanatory-quality test. It is the *explanatory quality* of a hypothesis, on its own but also taken in comparison to others, which contributes essentially to the warrant for its acceptability. The advantage of proceeding the way suggested is that there is no longer (great) pressure to answer in the abstract the question about the link between loveliness and likeliness. It should be implicit in the abstraction that, if anything, the context of application will settle the question: how likely is the best explanation?

Seen in this light, IBE should be considered as an inferential *genus*. The several species of the genus IBE are distinguished, among other things, by plugging assorted conceptions of explanation in the reasoning schema that constitutes the genus. For instance, if the relevant notion of explanation is causal, IBE becomes an inference to the best causal explanation. Or, if the relevant notion of explanation is subsumption under laws, IBE becomes a kind of inference to the best nomological explanation, and so forth. This might be taken to be an *ad hoc* attitude to the seemingly pressing issue of the model of explanation that features in IBE. But it is not. Many things can be explanations and the explanatory relation is not fixed and immutable. Attempts to cast the concept of explanation under an abstract model are parts of the search for the Holy Grail. For one, there is no extant model of explanation that covers all aspects of the explanatory relation. It is not implausible to think that 'explanation' is a cluster-concept, which brings together many distinct but resembling explanatory relations. For another, IBE should not be seen as issuing commitment to a single explanation. Rather, the commitment is to one (the best) among *competing* explanations. An explanandum might admit of different *types* of explanation (causal, nomological, mechanistic, unificatory etc.) There is no reason to take one of them as being the right type of explanation. In many cases, different types of explanation will be compatible with each other. Which of them will be preferred will depend on interests and on the details of the case at hand.

Since my book on scientific realism was published a number of philosophers of science have taken issue with its arguments and have raised serious worries and counter-arguments.[2] I have dealt with some of them in the present book. But I still owe replies to a number of others. The fact is that in the last few years my own thinking on realism has been shaped by a more careful and close reading of the writings of some logical positivists/empiricists on scientific realism – especially those of Moritz Schlick's, Hans Reichenbach's and Herbert Feigl's. Their attempt to carve a space for empiricists to be realists and to sail clear both from the Schylla of metaphysics (of the pre-Kantian form) and the Charybdis of instrumentalism makes a lot of sense to me. My current thinking has also been influenced by neo-Humean approaches to causation – where regularities rule. It remains to be seen how this dual empiricist impact on my thinking will affect my realist commitments.

Platania and Athens
August–September 2008

# Part I

# Scientific Realism

# 1
# The Present State of the Scientific Realism Debate

> The unique attraction of realism is the nice balance of feasibility and dignity that it offers to our quest of knowledge. (...) We want the mountain to be climbable, but we also want it to be a real mountain, not some sort of reification of aspects of ourselves.
>
> Crispin Wright (1988, 25)

Once upon a time there was a feeling in the philosophy of science community that the scientific realism debate had run out of steam. Arthur Fine went as far as to declare that 'realism is well and truly dead' (1986a, 112) and to compose the obituary of the debate, *aka* the Natural Ontological Attitude. Fortunately, the allegations of premature death failed to persuade many philosophers, for whom the scientific realism debate has had a glorious past and a very promising future. In the last dozen of years only there have been a number of books which cast a fresh eye over the issue of scientific realism, such as those by Suppe (1989), Putnam (1990), Almeder (1992), Wright (1992), Kitcher (1993a), Aronson, Harré & Way (1994), Brown (1994), Laudan (1996), Leplin (1997), Kukla (1998), Trout (1998), Cartwright (1999), Giere (1999), Niiniluoto (1999) and Psillos (1999). Although these books differ vastly in their approaches and in their substantive theses, they can all be seen as participating in a common project: to characterise carefully the main features of the realism debate and to offer new ways of either exploring old arguments or thinking in novel terms about the debate itself.

In this chapter I discuss the present state of the scientific realism debate with an eye to important but hitherto unexplored suggestions and open issues that need further work.

## 1.1 Modesty and presumptuousness

In Psillos (1999), I offered the following three theses as constitutive of scientific realism. Each of these is meant to warn off a certain non-realist approach.

*The Metaphysical Thesis*: The world has a definite and mind-independent structure.

*The Semantic Thesis*: Scientific theories should be taken at face value. They are truth-conditioned descriptions of their intended domain, both observable and unobservable. Hence, they are capable of being true or false. The theoretical terms featuring in theories have putative factual reference. So, if scientific theories are true, the unobservable entities they posit populate the world.

*The Epistemic Thesis*: Mature and predictively successful scientific theories *are* well confirmed and approximately true of the world. So, the entities posited by them, or, at any rate, entities very similar to those posited, inhabit the world.

The *first* thesis means to make scientific realism distinct from all those anti-realist accounts of science, be they traditional idealist and phenomenalist or the more modern verificationist accounts of Michael Dummett's (1982), and Hilary Putnam's (1981, 1990) which, based on an epistemic understanding of the concept of truth, allow no divergence between what there is in the world and what is licensed as existing by a suitable set of epistemic practices, norms and conditions. It implies that if the unobservable natural kinds posited by theories exist at all, they exist independently of the scientists' ability to be in a position to know, verify, recognise, and the cognate, that they do.

The *second* thesis – semantic realism – renders scientific realism different from *eliminative instrumentalist* and *reductive empiricist* accounts. Eliminative instrumentalism (most notably in the form associated with Craig's Theorem) takes the 'cash value' of scientific theories to be fully captured by what theories say about the observable world. This position typically treats theoretical claims as syntactic-mathematical constructs which lack truth-conditions, and hence any assertoric content. Reductive empiricism treats theoretical discourse as being disguised talk about observables and their actual (and possible) behaviour. It is consistent with the claim that theoretical assertions have truth-values, but understands their truth-conditions *reductively*: they are fully captured in an observational vocabulary. Opposing these two positions, scientific realism is an 'ontologically inflationary' view. Understood realistically, the theory admits of a literal interpretation, namely, an interpretation in which the world is (or, at least, can be) populated by a host of unobservable entities and processes.

The *third* thesis – epistemic optimism – is meant to distinguish scientific realism from *agnostic* or *sceptical* versions of empiricism (cf. van Fraassen 1980, 1985). Its thrust is that science *can* and *does* deliver theoretical truth[1] no less than it can and does deliver observational truth. It's an implicit part of the realist thesis that the ampliative–abductive methods employed by scientists to arrive at their theoretical beliefs are reliable: they tend to generate approximately true beliefs and theories.

Semantic realism is not contested any more. Theoretical discourse is taken to be irreducible and assertoric (contentful) by all sides of the debate. Making semantic realism the object of philosophical consensus was by no means an easy feat, since it involved two highly non-trivial philosophical moves: *first*, the liberalisation of empiricism with its concomitant admission that theoretical discourse has 'excess content', that is, content that cannot be fully captured by means of paraphrase into observational discourse; and *second*, a battery of indispensability arguments which suggested that theoretical terms are indispensable for any attempt to arrive, in Rudolf Carnap's (1939, 64) words, at 'a powerful and efficacious system of laws' and to establish an inductive systematisation of empirical laws (cf. Hempel 1958).

Given this, the distinctive of scientific realism is that it makes two claims in tandem, one of which (to explore Wright's (1992, 1–2) terminology) is 'modest', while the other is more 'presumptuous'. The *modest* claim is that there is an independent and largely unobservable-by-means-of-the-senses world that science tries to map. The *presumptuous* claim is that although this world is independent of human cognitive activity, science succeeds in arriving at a more or less faithful representation of it, that is of knowing the truth (or at least *some* truth) about it.

For many philosophers, this is *ab initio* an impossible combination of views: if the world is independent of our abilities or capacities to investigate it and to recognise the truth of our theories of it, how can it possibly be knowable? Two options then appear to be open to prospective realists: either to compromise presumptuousness or to compromise modesty.

### 1.1.1 Compromising presumptuousness

Here the cue is taken from Karl Popper's (1982). Take realism to be a thesis about the aim of science (truth), leaving entirely open the issue of whether this aim is (or can ever be) achieved. Implicit in this strand is that truth is understood realist-style (in the sense of correspondence with the world) in order not to compromise modesty as well. Popper made the headlines by claiming that scientists can never say that this aim has been achieved, but that truth is somehow magically approached by the increasing verisimilitude of successive theories; *magically* because there is nothing in Popper's account of verisimilitude which, even if it worked,[2] guarantees that there is a 'march on truth'. As we shall see in Chapter 3, Musgrave (1996, 23) agrees with Popper that realism is primarily an *axiological thesis*: science aims for true theories. There is clear motivation for this compromise: even if all theories scientists ever come up with are false, realism isn't threatened. Musgrave doesn't think that all scientific theories have been or will be outright false. But he does take this issue (whatever its outcome may be) to have no bearing on whether realism is a correct attitude to science.[3] There are, however, inevitable philosophical worries about the axiological characterisation to realism. First, it seems rather vacuous. Realism is rendered immune to any serious criticism that stems from the empirical claim that science has a poor record in truth-tracking (cf. Laudan 1984). Second, aiming at a goal (truth) whose achievability by the scientific method is left unspecified makes its supposed regulative role totally mysterious. Finally, all the excitement of the realist claim that science engages in a cognitive activity that pushes back the frontiers of ignorance and error is lost.

It seems irresistible that the only real option available for presumptuous realists is to compromise their modesty: if the world isn't in any heavyweight way independent of us, its knowability can be safeguarded. Compromising modesty is typically effected by coupling realism with an epistemic notion of truth which *guarantees* that the truth does not lie outside our cognitive scope.

### 1.1.2 Compromising modesty

Here the main cue is taken from Putnam's (1981). Take realism to involve an epistemic conception of truth, that is, a conception of truth which guarantees that there cannot be a divergence between what an ideal science will assert of the world and what happens (or there is) in the world. This line has been exploited by Brian Ellis (1985) and Nicholas Jardine (1986). For Ellis, truth is 'what we should believe, if our knowledge were perfected, if it were based on total evidence, was internally coherent and was theoretically integrated in the best possible way' (1985, 68). There are many problems with this view that I won't rehearse here (cf. Newton-Smith 1989b; Psillos 1999, 253–5). The only thing to note is that it's not obvious at all whether the suggested theory of truth is stable. To use Jardine's (1986, 35) words, the needed concept of truth should be neither too 'secular', nor too 'theological'. It should avoid an awkward dependence of truth on the vagaries of our evolving epistemic values, but it should link truth to *some* notion of not-too-inaccessible epistemic justification. In the attempt to break away from 'secular' notions of truth and to make truth a standing and stable property, it moves towards a 'theological' notion: the justification procedures become so ideal that they lose any intended connection with humanly realisable conditions. In the end, the required epistemic conception of truth becomes either 'secular', resulting is an implausible relativism, or 'theological', and hence not so radically different from a (realist) non-epistemic understanding of truth, according to which truth *outruns* the possibility of (even ideal-limit) justification. To be sure, Putnam (1990, viii) has dissociated his views on truth from the (Peircean) ideal-limit theory on the grounds that the latter is 'fantastic (or utopian)'. Still, his proposed alternative which ties ascriptions of truth with the exemplification of 'sufficiently good epistemic situations' fares no better than the Peircean theory vis-à-vis the secular/theological test above. One can always ask: what other than the realist-style truth of a proposition can *guarantee* that the sufficiently good conditions of justification obtain?[4]

There is an interesting dual thesis advanced by Crispin Wright that (a) a suitable epistemic concept of truth does *not* necessarily compromise the modesty of scientific realism and (b) the best hope for the presumptuousness of scientific realism rests on a broadly verificationist (epistemic) understanding of truth. For Wright, scientific realism stands mainly for (a) anti-reductionism and (b) the claim that theoretical discourse is apt for 'representation or fit with objective worldly states of affairs' (1992, 159). The first part of his thesis stems from the thought that the anti-reductionist stance of semantic realism, which treats theoretical discourse as apt for representation, is consistent with a (suitably) 'evidentially constrained' account of truth. This is so because, he claims, scientific realists may accept *both* that theoretical assertions faithfully represent worldly states-of-affairs *and* that these states-of-affairs are 'in principle' detectable (and hence, in principle verifiable). In particular, these worldly states-of-affairs need not be representable in a humanly intelligible way. On this view, the world ends up being independent of human representation (as scientific realism requires), yet the world is in principle detectable, which means that the relevant notion of truth is suitably 'evidentially constrained', and hence epistemic. (The motto for Wright's verificationist scientific realists would be: there is no in principle undetectable truth.) The second part of Wright's thesis stems from the thought that the realists' epistemic optimism requires that 'the harvest of best methods is (likely to be) truth and may, *qua* so harvested, be reasonably so regarded (1986, 262). But, he goes on, if truth is not taken to be what is 'essentially certifiable by best method' (as a verificationist realist would have it), there is no guarantee that truth is achievable. So, Wright concludes, either the door is left open to a van Fraassen-type sceptic, or to a Quinean pragmatist who 'cashes out' talk of truth is terms of talk about a(n) (everlasting) set of simplicity-guided adjustments in our evolving network of beliefs in response to empirical anomalies.[5]

Wright (1992) has presented a 'minimalist' concept of truth (not to be confused with Paul Horwich's (1990) account) which is characterised by some 'syntactic and semantic platitudes' (e.g., Tarski's T-schema, good behaviour with respect to negation, a 'thin' correspondence intuition, stability and others). Satisfaction of these platitudes (on Wright's proposal) guarantees that a certain discourse with a truth-predicate in it is assertoric (apt for truth and falsity), but leaves open the question whether the concept of truth has a more robust substance. Some realists believe that the concept of truth does have this more robust (non-epistemic) substance which is captured by a 'thick' notion of correspondence with reality, namely, that the *source* of the truth of theoretical assertions is worldly states-of-affair. This notion is taken by realists to be epistemically unconstrained. Wright juxtaposes to this realist notion of truth an epistemically constrained one: 'superassertibility' (1992, 48). He takes it to be the case that superassertibility meets the minimalist requirements noted above, and then asks whether there are features of a discourse which dictate that this discourse needs or implicates a concept of truth *stronger than* superassertibility. He proposes four criteria for judging whether a discourse implicates an epistemic or a non-epistemic conception of truth (over and above the minimalist common ground): extensional divergence, convergence of opinion (or Cognitive Command), the Euthyphro Contrast, and the width of cosmological role. Put in a nutshell, Wright's claim is the following. It may be that truth (realist-style) and superassertibility are extensionally divergent notions (there are truths which are not superassertible and/or conversely). It may be that truth (realist-style) features in best explanations of why there is convergence-of-opinion in a discourse. It may be that the direction of dependence between truth (realist-style) and superassertibility is one-way only: it's because certain statements are true (realist-style) that they are superassertible and not conversely. And it may be that the statements in a discourse play a wide cosmological role in that their truth (realist-style) contributes to the explanation of assertions and attitudes in other spheres or discourses. This (extremely compact presentation of Wright's seminal idea) leads me to the following conjecture. Even if Wright is right in pointing out that, *prima facie*, scientific realists need not compromise their modesty by adopting an epistemically constrained conception of truth, the very features of the truth-predicate implicated in the assertoric theoretical discourse in science are such that it satisfies all criteria that Wright himself has suggested as pointing towards the operation (or implication) of a (realist-style) concept of truth in a discourse. If this conjecture is right, then the realist aspiration to modesty *ipso facto* implicates a substantive non-epistemic conception of truth.

What about the second part of Wright's thesis, namely, that scientific realists had better adopt an epistemic conception of truth if they are to retain their epistemic optimism? The problem with this suggestion (which Wright recognises and tries to meet) is that a verificationist version of scientific realism brings with it all of the problems that discredited verificationism as a philosophical theory of meaning (and truth). In particular, the viability of Wright's second thesis depends on two premises: first, that radical underdetermination of theories by evidence is a priori impossible; second, that we can make sense of an observation language which is theory-free and which is used to 'cash out' the suitable notion of verifiability. As for the first premise, it seems obvious that the very logical possibility of two or more mutually incompatible theories being empirically equivalent entails (on the assumption that only one of them can be true) that truth doesn't necessarily lie within our cognitive capacities and practices. As for the second premise, if observation is theory-loaded in such a way that we cannot segregate a set of theory-neutral 'observation reports', we cannot even begin to formulate the thesis that theoretical assertions are true in the sense that they are fully verifiable by means of 'observation reports'.

Some realists (e.g., Michael Devitt and Horwich) take scientific realism to be an ontological doctrine which asserts the existence of unobservable entities, and claim that no doctrine of truth is constitutive of realism. Here company is parted, however. Devitt (1984, Chapter 4) argues that insofar as a concept of truth is involved in the *defence* of realism, it should be a correspondence account in order to safeguard that the world is independent in its existence and nature from what we believe. Horwich (1997), on the other hand, after declaring that the scientific realism debate is about the independence and accessibility of facts about unobservable entities, takes the view that a 'deflationary' conception of truth (which is itself lightweight and metaphysically neutral) is all that is needed for the defence of scientific realism. His core thought is that the truth-predicate doesn't stand for any complex property, but is a quasi-logical device for forming generalisations over propositions.

One can of course pit Devitt's defence of correspondence truth against Horwich's deflationism. But the serious philosophical issue that remains is Horwich's (1997) thesis that the scientific realism debate can be fully stated and explained without any substantive (i.e., non deflationary) concept of truth. In particular, Horwich claims that even when the concept of truth is explicitly mentioned in a realist (or anti-realist) thesis, for example, when realists say that science achieves theoretical truth, or when instrumentalists say that theoretical hypotheses are truth-valueless, or when verificationists say that all truths are verifiable, even *then* it can be captured by a deflationist understanding of truth. But I doubt that this is so easily established. When realists say, for instance, that theoretical discourse should be understood literally, they imply that theoretical assertions shouldn't be taken to be translatable into a vocabulary which commits only to observable states-of-affair. The notion of translatability (or its lack) may inevitably involve reference to sameness (difference) of truth-conditions, which, arguably, are not part of the resources available to the deflationist (cf. Field 1992, 324–5).[6]

### 1.1.3 Conceptual independence and epistemic luck

Despite attempts to force a compromise on scientific realists, neither modesty nor presumptuousness has to go. From the claim of independence of the world from human cognitive activity it does not follow either that human inquirers are cognitively closed to this world or that when they come to know it, they somehow *constitute* it as the object of their investigation. All depends on how exactly we understand the realist claim of mind-independence. As will be explained in Chapter 2, Section 2.4, it should be taken to assert the logical-conceptual independence of the world: there is no conceptual or logical link between the truth of a statement and our ability to recognise it, assert it, superassert it and the like. The entities that science studies and finds truths about are deemed independent of us (or of mind in general) not in any causal sense, but only in a logical sense: they are not the outcome of, nor are they constituted by (whatever that means), our conceptualisations and theorising. This kind of independence is consistent with the claim that science and its methodology are causally dependent on the world. In fact, the realists' claim that the scientists' methods of interaction with the world are such that, at least in favourable circumstances, can lead to the formation of warranted beliefs about the 'deep structure' of the world presupposes *causal contact* with the world.

Despite several pages of philosophical argumentation that this contact with the independent world is impossible because it would amount to 'getting out of our skin' (cf. Rorty 1991, 46ff), or because it's 'conceptually contaminated' (Fine 1986b, 151), it's a simple truth that our (inevitably) conceptual give-and-take with the world need not lead to the neo-idealist (or neo-Kantian) thought that the causal structure of the world is a reflection (or projection) of our concepts and intuitions. The independence of the world needn't be compromised. And it cannot be compromised unless one adopts the implausible view that the worldly entities are *causally constituted* as entities by our conceptual and epistemic conditions, capacities and practices. To be sure, realists need to grant that their 'epistemic optimism' that science has succeeded in tracking truth requires a *epistemic luck*: it's not a priori true that science has to be successful in truth-tracking. If science does succeed in truth-tracking, this is a *contingent fact* about the way the world is and the way scientific method and theories have managed to 'latch onto' it (cf. Boyd 1981). Accordingly, the presumptuousness of realism is a contingent thesis that needs to (and can) be supported and explained by argument which shows that the ampliative–abductive methods of science can produce theoretical truths about the world and deliver theoretical knowledge.

If neither modesty nor presumptuousness need compromising, isn't there still an issue as to how presumptuous scientific realism should be? I think we should reflect a bit on what exactly the *philosophical* problem is. I take it to be the following: is there any good reason to believe that science cannot achieve theoretical truth? That is, is there any good reason to believe that after we have understood the theoretical statements of scientific theories as expressing genuine propositions, we can never be in a warranted position to claim that they are true (or at least, more likely to be true than false)? There are some subtle issues here (to which we shall return below and in Chapter 2), but once we view the problem as just suggested, it transpires that what realism should imply by its presumptuousness is that theoretical truth is achievable (and knowable) no less than observational truth.

This last claim may have a *thin* and a *thick* version. The thin version is defended by Jarrett Leplin (1997). His 'Minimal Epistemic Realism' is the thesis that 'there are possible empirical conditions that would warrant attributing some measure of truth to theories – not merely to their observable consequences, but to theories themselves' (op. cit., 102). As Leplin is aware, many realists would go for a thicker version. This version should imply (and be engaged in the defence of the claim) that the ampliative–abductive methods of science are reliable and do confer justification on theoretical assertions. This thick version is the standing result of Richard Boyd's contribution to the defence of realism. But why do we need it? A 'thin' account cannot issue in rational or warranted belief in the unobservable entities posited by science (and the assertions made about them). All the thin claim asserts is a subjunctive connection between some possible empirical conditions and the truth of some theoretical assertions. This cannot be the litmus test for scientific realism because, suitably understood, it's universally acknowledged as possible. Not only are we given *no* guarantee that this possible connection may be actual (a condition required for the belief in the truth of a theoretical assertion). More importantly, any attempt to offer such a guarantee would have to go via a defence of the method that connects some empirical condition with the truth of a theoretical assertion. Hence, the defence of the rationality and reliability of these methods cannot be eschewed. To me, all this means that the presumptuous strand in the realist thought should be thick. One issue that needs to be explored – as hinted at by Fred Suppe (1989, 340–6) and developed by Kitcher (1993a, Chapter 3) – is how standard epistemological theories of justification, reliability and belief formation can be employed in the realism debate. It may turn out, as I (Psillos 1999, 83–6) and Suppe (1989, 352) believe it does, that the debate on scientific realism is best conducted in the context of broader epistemological theories about the nature of knowledge, justification etc. (This is issue is explored in some detail in Chapter 10.)

So far, we have resisted the claim that the concept of truth involved in scientific realism should be something less than a 'correspondence with reality'. The well-known pressures have led some realists to back down (e.g., Giere 1999, 6). Others, however, have tried to explicate the notion of correspondence in such a way as to remove from it any sense in which it is 'metaphysically mysterious'. Of these attempts, Kitcher's (1993b, 167–9) stands out because he shows that this notion (a) need not commit us to an implausible view that we should (or need to) compare our assertions with the world and (b) arises out of the idea that a fit between representations and reality *explains* patterns of successful action and intervention. A correspondence account of truth is just another theory which can be judged for (and accepted on the basis of) its explanatory merits.

## 1.2 Epistemic optimism

It's hard to exaggerate the role that Sellars played in the realist turn during the 1960s. His attack on the 'myth of the given' and his endorsement of the 'scientific image', according to which what is real is what successful scientific theories posit, prioritised scientific theories over folk theories of the 'manifest image' as our guide to what there is (cf. Churchland 1979). (More details about Sellars's arguments for scientific realism are given in Chapter 5.)

Jack Smart (1963, 39) and Maxwell (1962, 18) followed suit by offering arguments for realism based on the explanation of the success of science. If all these unobservable entities don't exist, if theoretical assertions are not well-confirmed and true descriptions of an unobservable world, it isn't possible to explain the empirical success of science and the predicted observed correlations among observable entities. Putnam (1975, 73) turned all this into a famous slogan: realism 'is the only philosophy of science that does not make the success of science a miracle'. Hence, the well-known 'no-miracles' argument for realism (NMA). To be sure, the central thought in this argument is that the realist assertions offer not the only but the *best* explanation of the success of science. Be that as it may, the point of NMA is that the success of scientific theories lends credence to the following two theses: (a) scientific theories should be interpreted realistically and (b) these theories, so interpreted, are well confirmed because they entail well-confirmed predictions. (A relatively detailed discussion of NMA is given in Chapter 3.)

The original authors of NMA didn't place emphasis on *novel* predictions, which are the litmus test for the ability of alternative approaches to science to explain the success of science. For only on a realist understanding, novel predictions about the phenomena *come as no surprise*. Yet, there has been notorious disagreement as to how exactly the conditions of novelty should be understood. A novel prediction has been taken to be the prediction of a phenomenon whose existence is ascertained only after a theory has predicted it. This, however, cannot be the whole story since theories get support also from explaining already known phenomena. It's been suggested (e.g., Worrall 1985) that the 'temporal view' of novelty should be replaced by a 'novelty-in-use' view: a prediction of an already known phenomenon can be use-novel with respect to some theory provided that information about this phenomenon was not *used* in the construction of the theory. Yet, it's been very difficult to make precise the intuitive idea of 'use novelty'. A fresh analysis comes from Leplin (1997, 77) who explicates 'novelty' by reference to two conditions: 'independence' and 'uniqueness'. The thrust is that a prediction of a phenomenon O is novel for a theory T if no information about O is necessary for the prediction of O by T, and if at the time T explains and predicts O, no other theory 'provides any viable reason to expect' O. If these requirements are satisfied, it's hard to see what other than the relevant truth of the theory T could best explain the novel predictions.[7]

Why has the realist turn come under so much pressure? The main target of the non-realist onslaught has been realism's epistemic optimism. Note that the original Smart–Maxwell formulation of the 'no miracle' argument rested on the assumption that once semantic realism is established, belief in the truth of genuinely successful scientific theories is (almost inevitably) rationally compelling. Van Fraassen's (1980) reaction to this was that the ampliative–abductive methodology of science fails to connect robustly empirical success and truth: two or more mutually incompatible theories can nonetheless be empirically congruent and hence equally empirically successful. Given that at most one of them can be true, semantic realism can still stand but be accompanied by a sceptical attitude towards the truth of scientific theories.

Realists face a dilemma. As W.H. Newton-Smith (1978, 88) pointed out, realists can cling on an 'Ignorance Response' or an 'Arrogance Response'. On the first horn, realists choose to hang on to a realist metaphysics of an independent world, but sacrifice their epistemic optimism. On the second horn, they try to secure some epistemic optimism, but sacrifice the independence of the world by endorsing a view which denies that there are 'inaccessible facts' which can make one of the many rival theories true. In a way, van Fraassen's own attitude amounts to the 'Ignorance Response'.[8] As for the 'Arrogance Response', it's hard to see how one can be a realist and still endorse it. Trimming down the content of the world so that it contains no inaccessible facts leaves three options available (all of which should be repugnant to realists). The first is to re-interpret the empirically equivalent theories so that they are not understood literally and the apparent conflict among them doesn't even arise (an option entertained by some Logical Empiricists). The second (as already noted in Section 1.2) is to adopt an epistemic notion of truth which makes it the case that only one of the empirically equivalent theories passes the truth-test (cf. Jardine

1986). And the third is to argue that all these theories are true, thereby relativising the concept of truth (cf. some time-slice of Quine 1975, 327–8).

Can realists eschew the 'Ignorance Response'? The gist of van Fraassen's challenge is that the explanatory virtues which are part and parcel of the ampliative–abductive methodology of science need not (and perhaps cannot) be taken to be truth-tropic. Hence, any realist hope to forgo the 'Ignorance Response' by grounding their epistemic optimism on explanatory considerations seems to vanish. Not so fast, though.

Putnam's standing contribution to the realist cause is his thought that the defence of realism cannot be a piece of a priori epistemology, but rather part and parcel of an empirical-naturalistic programme which claims that realism is the best empirical hypothesis of the success of science. Capitalising on this thought, Boyd (1981, 1984) embarked on an attempt to establish the accessibility of (and rational belief in) theoretical truth by trying to defend the reliability of ampliative–abductive inferences. This well-known *abductive defence of realism* starts from the fact that the heavily theory-laden scientific methodology is instrumentally reliable (i.e., it yields correct predictions and is empirically successful) and argues that the best explanation of this instrumental reliability is that the background theories (which inform and dictate the methods used by scientists) are approximately true. This is a philosophical (second order) inference to the best explanation (IBE) which suggests that there is a contingent (a posteriori) link between ampliative–abductive methodology (and the concomitant notion of 'best explanation') and truth. It is this argument that grounds the realists' epistemic optimism. It also removes the sting from the rival argument from the underdetermination of theories by evidence (UTE). For two empirically equivalent theories may not be (as a matter of contingent fact) equally good in their explanatory virtues. Hence one of them may well be the best explanation of the evidence and command rational belief.

In any case, UTE rests on two questionable premises: (I) For any theory T there is at least another one incompatible theory T′ which is empirically congruent with T. (II) If two theories are empirically equivalent, they are epistemically equivalent too (i.e., equally confirmed or supported by the evidence). Both premises have been forcefully challenged by realists. Some (e.g., Newton-Smith 1987) have challenged (I) on the grounds that the thesis it encapsulates is not proven. Note, in passing, that realism should be happy with local scepticism. It may turn out that some domains of inquiry (e.g., the deep structure of space-time) are beyond our ken. Others (e.g., Glymour 1980; Boyd 1981; Laudan & Leplin 1991; Laudan 1996; Psillos 1999) have objected to (II). Here there are, on the face of it, two strategies available. One (IIa) is to argue that even if we take only empirical evidence in the strictest sense as bearing on the epistemic support of the theory, it does not follow that the class of the observational consequences of the theory is co-extensional with the class of empirical facts that can lend support to the theory. An obvious counter-example to the claim of co-extensionality is that a theory can get indirect support by evidence it does not directly entail. The other strategy (IIb) is to note that theoretical virtues are epistemic in character and hence can bear on the support of the theory. Here again there are two options available to realists: (IIb.1) to argue that some theoretical virtues are constitutive marks of truth (e.g., McMullin 1987); or (IIb.2) to argue for a broad conception of evidence which takes the theoretical virtues to be broadly empirical and contingent marks of truth (cf. Boyd 1981; Churchland 1985; Lycan 1988). (IIb.2) is an attractive strategy for two reasons: (a) it challenges the strict empiricist conception of evidence and its relation to rational belief; (b) it removes the apparent tension between modesty and presumptuousness, without also forging an a priori link between theoretical virtues and truth. (IIb.2) is perhaps the most difficult position to defend, but on closer inspection it may well turn out that (IIa) and (IIb.2) are, at root, the very same strategy (cf. Psillos 1999, 171–6).[9]

Not all defenders of realism take the abductive defence of IBE to be central in the defence of realism. There are a few specific problems here and one more general philosophical. The specific problems regard the notion of explanation and the conditions under which it can be balled 'best'. Some realists countenance specific forms of causal explanation (e.g., Salmon (1984) for the so-called common cause principle, or Cartwright (1983) for 'inference to the most probable cause') but deny that they can suitably generalise to engender a blanket notion of IBE. Others (e.g., Lipton 1991) try to provide (descriptively) an account of when a (potential) explanation is best and then to tell a story as to when this explanation licences inference. In the same boat, Ilkka Niiniluoto (1999, 185–92) sketches a formal model of IBE in which the 'best explanation' is linked to the 'best confirmed' hypothesis, given the evidence. Finally, there are those (e.g., Miller 1987) who argue that there cannot be a general mode of inference called IBE, but instead that local ampliative inferences in science are licensed only when they are backed up by 'topic-specific truisms', that is principles which are so entrenched that no-one in the specific domain can seriously deny them. This last position, however, is sensitive to the issue of what renders these principles 'truisms' if not the fact that they have been arrived at by a legitimate application of IBE. (For more on these matters see Chapter 10. My current views on the structure of IBE are sketched in Psillos 2007a).

What I called the general philosophical problem of the abductive defence of realism has caused a heated discussion. It has been argued (cf. Laudan 1984, 134; van Fraassen 1985, 255; Fine 1986a,b) that the realists' *use* of (a second-order) IBE in defence of realism is circular and question-begging. For, the thought is, it takes for granted the reliability of a mode of inference which is doubted by non-realists. This challenge has led some realists to question the viability of the abductive strategy. Newton-Smith (1989a, 179), for instance, called the realism associated with this strategy 'faded'. And Rom Harré (1988) left behind 'truth realism' and its 'deeply flawed' abductive defence in favour of a methodological strategy, which he called 'policy realism' (cf. also Hendry 1995).

This issue is a focal point of the debate. A proper appreciation of what is at stake presupposes a better understanding of the broader epistemological agendas of the participants. As is explained in detail in Psillos (1999, Chapter 4), the abductive defence of realism proceeds within a broad naturalistic framework in which the charge of circularity loses its bite because what is sought is not justification of inferential methods and practices (at least in the neo-Cartesian internalist sense) but their explanation and defence (in the epistemological externalist sense). It's not as if the abductive defence of realism should persuade a committed opponent of realism to change sides. Strict empiricists, for instance, are not likely to be moved by any defence of IBE, be it circular or straight, precisely because as Ernan McMullin (1994, 100) has noted, they simply choose to tolerate unexplained regularities and phenomena. (One such regularity is that science has been instrumentally reliable and successful.) Van Fraassen's insistence that the explanatory virtues are merely pragmatic is just a further twist to this tolerance to the unexplained. So, strict empiricists deny the abductive defence of realism not so much because it's circular (they would deny a defence of IBE even if it was straight), but mainly because they refrain from accepting the existence of unobservable entities on any grounds that transcend what can be derived from naked-eye observations. But unless this attitude is itself the most reasonable to adopt (something that I doubt), it doesn't follow that IBE is unreasonable.[10] Nor does it follow that the employment of IBE in an abductive defence of the reliability of IBE is question-begging and unpersuasive. Many (if not all) use modus ponens unreflectively as a sound inferential rule and yet an establishment of the soundness of modus ponens at the meta-level by an argument which effectively uses modus ponens can still explain to them why and in virtue of what features deductive reasoning is sound. In any case, realists vary in the extent to which they adopt the abductive defence of the reliability of IBE. There are those brazen realists, like Boyd, Trout (1998) and myself (Psillos 1999) who take the charge of circularity seriously and try to meet it within a naturalistic perspective. One central thought in this camp is that there is just abduction as the general mode of ampliative reasoning and if this fails, then no ampliative reasoning (and hence no learning from experience) is possible (for more details on this, see Chapter 10). There are the temperate realists (cf. Leplin 1997, 116) who capitalise on the thought that abduction and induction are distinct modes of reasoning and try to argue that IBE is no worse than ordinary inductions which are OK for non-realists. Finally, there are realists (like Brown 1994, Chapter 1) who side-step the charge of circularity and argue that the explanatory story told by realism is just more adventurous and enlightening than alternative stories.

Yet, there is a deep empirical challenge to realism and its abductive defence: the Pessimistic Induction. As Larry Laudan (1984) has pointed out, the history of science is replete with theories that were once considered to be empirically successful and fruitful, but which turned out to be false and were abandoned. If the history of science is the wasteland of aborted 'best theoretical explanations' of the evidence, then it might well be that current best explanatory theories might take the route to this wasteland in due course. Not all realists find this argument threatening. Some (e.g., Devitt 1984) find it simply (and correctly) overstated. Others (e.g., Almeder 1992) take a 'blind realist' stance: at any given stage of inquiry some of our theoretical beliefs are true, yet we can never tell which are such because 'we have no reliable way of determining which of our currently completely authorised beliefs will suffer truth-value revision in the future' (Almeder 1992, 178).[11] What about those of us who think that we should take seriously the Pessimistic Induction and try to meet it?

Although other strategies may be available, the best defence of realism is to try to reconcile the historical record with some form of realism. In order to do this, realists should be more selective in what they are realists about. A claim that now emerges with some force is that theory-change is not as radical and discontinuous as the opponents of scientific realism have suggested. Realists have showed that there are ways to identify the theoretical constituents of abandoned scientific theories which essentially contributed to their successes, to separate them from others that were 'idle' – or as Kitcher has put it, merely 'presuppositional posits' – and to demonstrate that the components that made essential contributions to the theory's empirical success were those retained in subsequent theories of the same domain (cf. Kitcher 1993a; Psillos 1999, Chapter 5). Given this, the fact that our current best theories may be replaced by others does not, necessarily, undermine scientific realism. All it shows is that (a) we cannot get at the truth all at once; and (b) our judgements from empirical support to approximate truth should be more refined and cautious in that they should only commit us to the theoretical constituents that do enjoy evidential support and contribute to the empirical successes of the theory. Realists ground their epistemic optimism on the fact that newer theories incorporate many theoretical constituents of their superseded predecessors, especially those constituents that have led to empirical successes. The substantive continuity in theory-change suggests that a rather stable network of theoretical principles and explanatory hypotheses has emerged, which has survived revolutionary changes, and has become part and parcel of our evolving scientific image of the world.

This reaction to the Pessimistic Induction has been initiated by Worrall's seminal (1989). What he called 'structural realism' is an attempt to capitalise on the fact that despite the radical changes at the theoretical level, successor theories have tended to retain the mathematical structure of their predecessors. Worrall's thought is that theories can successfully represent the structure of the world, although they tend to be wrong in their claims about the entities they posit. As we shall see in detail in the second part of the book, it turns out that this particular position is very difficult to defend (cf. Ladyman 1998; Psillos 1999, Chapter 7). Cartwright (1999, 4) has taken a different path. She is happy to go from the 'impressive empirical successes of our best physics theories' to 'the truth of these theories', but she denies that the assertions made by these theories are universal in scope. Rather, she goes for a 'local realism about a variety of different kinds of knowledge in a variety of different domains across a range of highly differentiated situations' (op. cit., 23) which tallies with her view that the world is best seen as disunified, with no laws or principles holding across the board and across different domains of inquiry. This is an issue that we shall discuss in detail in Chapter 6. Arguing as Cartwright does, for local truths which may vary from one model to another and from one domain to another, may involve a perspectival notion of truth with characteristics not suitable for realism.

Realists talk of approximate truth and take science and its methods to issue in approximately true beliefs. How much of a substantive concession this is is a matter of dispute. Laudan (1984) claims that the realist cause is doomed unless a formal semantic for approximate truth is in the offing. Ron Giere (1988) concedes this but claims that realists can do well with a notion of similarity between the theoretical model and the domain to which it applies. Aronson, Harré and Way (1994) try to make good on the notion of similarity by devising an informal account of approximate truth which rests on the view that theories are type-hierarchies of natural kinds. Others (e.g., Niiniluoto 1999) still think that there are good prospects for a formal (and consistent) explication of approximate truth. My own view (cf. Psillos 1999, Chapter 11) has been that we shouldn't be deterred in our philosophical disputes by formal issues if the operative notions are intuitively clear and do not lead to paradoxes. As Peter Smith (1998) has suggested, the intuitive notion of 'approximate truth' can be explicated sufficiently well to be usable along the following lines: for a statement P, 'P' is approximately true iff approximately P. This relegates much to the concept of approximation, but there is no reason to think that a domain-specific understanding of approximation is not robust enough to warrant ascription of approximate truth in statements about each domain.

Although, as we have seen, there have been extremely important and profound challenges to realism, the only articulated rival philosophical position that has emerged is van Fraassen's (1980) *Constructive Empiricism*.[12] This view is already familiar to everyone and has been thoroughly debated in Paul Churchland and Clifford Hooker (1985). Its core point is that committed empiricists cannot be forced to be scientific realists because (a) they can offer an alternative account of science which takes science to aim at empirical adequacy and involves only belief in the empirical adequacy of theories; and (b) this account of science is complete in the sense that there are no features of science and its practice which cannot be accounted for (or explained away) from this empiricist perspective. Given that it is impossible to do justice to the massive literature on this subject in the present space (but cf. Rosen 1994 and Psillos 1999, Chapter 9), I shall only make a general comment on the spirit of van Fraassen's approach.[13]

As Richard Miller (1987, 369) nicely put it, van Fraassen's stance is a kind of modern 'principle of tolerance'. Although van Fraassen

(1980) can be easily interpreted as trying to show that scientific realism is an irrational attitude (and hence that constructive empiricism is the only rational attitude to science), in his later work (van Fraassen 1989, 1994, 2000a) he emphasises a new conception of rationality according to which constructive empiricism is *no less* rational than scientific realism. This new conception of rationality suggests that 'what is rational to believe includes anything that one is not rationally compelled to disbelieve' (1989, 171–2). Hence, van Fraassen says, since scientific realism is not rationally compelling, and since disbelief in constructive empiricism is not rationally compelling either, constructive empiricism is an equally rational option. All this may suggest that the door to scepticism is open, since from the fact that one is not rationally compelled to disbelieve P, it doesn't follow that one has (or could possibly have) good reasons to believe P. But van Fraassen (1989, 178) feels no threat here for he *denies* the 'sceptical' claim that 'it is irrational to maintain unjustified opinion'. This new aspect of van Fraassen's philosophy and his post-1990 attempt to articulate empiricism have not yet received the attention they deserve.[14] As an attempt to initiate this discussion, it might be possible to argue that there are tensions in van Fraassen's account of rationality. In particular, one could argue that from the fact that scientific realism is not rationally compelling it doesn't follow that constructive empiricism is no less rational an option. (Compare: from the fact that it's not rationally compelling to believe in Darwinism it does not follow that belief in Creationism is equally rational.) In order, however, to substantiate this tension, we need to show at least one of the following things. Either that there are aspects of the phenomenology of science which do not make good sense under Constructive Empiricism – for example, I think (Psillos 1999, 204) that the practice of diachronic conjunction of theories offers such a test-case. Or, that the joint belief in the existence of observable middle-sized material objects *and* unobservables is more rational than the combination of belief in middle-sized objects and agnosticism about unobservables. This last thought has been explored by Peter Forrest (1994). It's motivated by the claim that belief in the existence of unobservable entities (as opposed to agnosticism about them) rests on the same grounds as belief in the existence of middle-sized material objects (as opposed to agnosticism about them). This last claim, however, presupposes that there is no principled difference between having reasons to believe in the existence of observables and having reasons to believe in the existence of unobservables. Despite van

Fraassen's insistence on the contrary, there is a lot of sound philosophical argument that the equation of the unobservable with the epistemically inaccessible is bankrupt (cf. Churchland 1985; Salmon 1985).

## *Addendum*

The preceding chapter aimed to offer a road map to the scientific realism debate. Since it was first published in 2000, a number of important books on scientific realism have appeared. Three of the most recent ones are Anjan Chakravartty's (2007), Derek Turner's (2007) and Christopher Norris's (2004) books. Taken together, these books present new challenges to realism and extend the debate to new territories. Here is some critical discussion of them.

### A1. Semirealism: a short critique

*A Metaphysics for Scientific Realism* aims to do two things.[15] The first is to develop a viable realist position which capitalises on insights offered by entity realism and structural realism, while transgressing them. Semirealism, as Chakravartty calls it, comes out as a form of selective scepticism which restricts commitment only to those parts of theories that can be interpreted as describing aspects of the world with which scientists have managed to be in causal contact. The second aim is to develop a metaphysical framework within which his semirealism can be cast. This is a non-Humean framework based on a dispositional account of properties and a network of *de re* necessities. Chakravartty admits that this is just one option available to scientific realists, but claims that it gives semirealism a high degree of internal coherence, and hence facilitates its defence.

These two aims create a somewhat unstable mix. If semirealism is the best hope for scientific realists and if it is seen as *requiring* commitment to a non-Humean metaphysical picture of the world, this might be reason enough to make scientific realism unattractive to all those who prefer barren metaphysical landscapes. Semirealism is so much metaphysically loaded that its very posture might be enough to give extra force to well-known empiricist arguments that tend to favour antirealism on the grounds that it alone can deliver us from metaphysics. If this rich metaphysical picture is an add-on to the selective epistemic commitments of semirealism (if scientific realists do not have to buy it, anyway), why not leaving it behind, thereby making scientific realism a more inclusive philosophical position?

Indeed, Chakravartty focuses on the empiricist critique of metaphysics (advanced recently by van Fraassen) and contrasts van Fraassen's stance empiricism with what Chakravartty calls 'the metaphysical stance' (being taken to be largely the stance of scientific realism). Given van Fraassen's own permissive conception of rationality, the metaphysical stance cannot be shown to be incoherent; hence it cannot be shown to be irrational. So, Chakravartty claims, the empiricist critique of metaphysics cannot win: it cannot block realism from incorporating a rich metaphysical outlook. But then again on Chakravartty's set-up, realism cannot win either. At best, there will be a tie between the empiricist stance and the metaphysical stance.

The motivation for semirealism comes from the usual suspect: the pessimistic induction. This suggests that epistemic commitment should be restricted to those parts of theories that are more likely to resist future revisions. Semirealism adopts the epistemic optimism of entity realism (which is grounded on cases of experimental manipulation of unobservable entities), but adds that knowledge of causal interactions presupposes knowledge of causal properties of particulars and relations between them. Semirealism also adopts the epistemic optimism of structural realism (which is based on structural invariance in theory-change), but adds that the operative notion of structure should be concrete and not abstract. Concrete causal structures consist in relations of first-order causal properties, which account for causal interactions. Chakravartty claims that these causal properties are best seen as being powers, as having a dispositional identity. Focusing on *concrete* causal structures (with or without the power-based account of properties) is a step forward. It implies that one cannot have knowledge of the structures without also having knowledge of the intrinsic natures of things that make up the structure. Thus, Chakravartty claims, knowledge of concrete causal structures contains 'unavoidably' knowledge of intrinsic natures of particulars, and vice versa. This is all fine. It reveals some of the problems associated with structural realism. But, then again, why bother to call these things concrete causal *structures*? This is a term of art, of course. But in the context of the current realism debate, it is meant to imply a contrast between a concrete relational system and its (abstract) structure; it also implies that there can be knowledge of structural characteristics without concomitant knowledge of non-structural characteristics (e.g., of the entities that instantiate a structure). (For more on this, see Chapter 8.) Within semirealism, concrete causal structures (and their knowledge) contain everything up to the very natures of particulars. Since nothing is left out, however, one cannot intelligibly talk about a substantial notion of *structure*.

Be that as it may, Chakravartty's key point is that the parts of theories to which realists should be epistemically committed should be those parts that can be interpreted as referring to a certain class of properties of concrete causal structures (or systems or whatever), namely, the 'detection' properties. These are properties that are causally detectable and in whose presence realists should most reasonably believe on the basis of the scientists' causal contact with the world. Detection properties are distinguished from auxiliary properties. These are attributed to particulars by theories but there is no reason to believe in their reality since they are not detected (though they might be detectable and become detected later on). Chakravartty appeals to what he calls a 'minimal interpretation' of the mathematical equations that make up a physical theory in order to demarcate the concrete causal structures associated with the detection properties from those associated with auxiliary ones. A minimal interpretation interprets realistically only those parts of equations that, in the context of a specific detection process, are indispensable for describing the (corresponding to that detection) concrete causal structures. Isn't there a tension here? If detection properties are specified independently of the theory, there is no need to interpret the theory *minimally* to get to them. If, however, they are specified in a theory-dependent way, this theory should already be interpreted prior to fixing the detection properties – and in all probability more than a minimal interpretation will be required to specify which properties are detection and which are auxiliaries.

As noted already, Chakravartty leaves the door open for less metaphysically loaded, but realist-friendly, conceptions of causation, laws and properties. His own view is that causation is a matter of continuous causal processes which are grounded in the dispositional nature of causal properties. Being powers, causal properties fix the laws in this and other worlds. They cast a net of *de re* necessities. This image of causal structuralism animates part of the book. Here is a worry, however. Semirealism has urged commitment to causally detectable properties and has clothed with suspicion all else (being merely auxiliary). But none of the extra stuff that Chakravartty (2007, 94) finds in the world (*de re* necessities, ungrounded dispositions and the like) are detectable. They are taken to be part of the baggage of semirealism because they play a certain explanatory role, notably they distinguish causal laws from merely accidental regularities (ibid., 94–5). But then, 'the deeper metaphysical foundations' of semirealism (ibid., 91) could well be (and in all probability are) mere auxiliaries, which can then be treated with the same suspicion as other scientific auxiliaries (like the ether). This creates another tension. We are invited to accept a certain set of double standards – one for scientific theories, and another for metaphysics. While in the case of scientific theories, epistemic optimism requires causal contact with the world, thus denying epistemic optimism based on the explanatory virtues of theories, in the case of the metaphysical foundations of semirealism, the only virtues on which one could base one's epistemic optimism are merely explanatory. Alternatively, if we allow explanatory considerations to play a role in science too (as distinct from mere causal detection), the very detection/auxiliary distinction that is so central in semirealism is put under a lot of strain. To put the point somewhat provocatively, the metaphysics of semirealism is the auxiliary system whose detection properties are Humean regularities and other metaphysically less fatty stuff.

### A2. The natural historical attitude: a short critique

Turner (2007) takes the past to be epistemogically problematic in two important senses – both of which involve an epistemic asymmetry between the past and the present. First, there is the asymmetry of manipulability, namely, an inability to intervene (to manipulate) the past.[16] Second, there is the asymmetry of the role of background theories, namely, that background theories about the past imply (as background theories about the present and the tiny do not) that a lot of evidence about the past has been irrevocably destroyed and a lot of possible information channels have been dampened. These asymmetries make rampant local underdetermination of theories about prehistory. In a great deal of cases, we are told, scientists face an issue of choice between rival but empirically equivalent *and* equally epistemically virtuous theories, with nothing in their hands (no means to generate new phenomena, no sources of additional information) to break the tie. There is a third asymmetry, Turner notes, the asymmetry of analogy, namely, that past posits seem to be analogous to current observable entities. This, it might be thought, eases the problem of knowing the past. But Turner argues that it is precisely this asymmetry that explains why scientists have made a number of mistakes about the past. Hence, relying on analogy is not a reliable way to learn about the past. The general conclusion drawn from the three asymmetries is that there are clear senses in which knowing the past is harder than knowing the tiny and hence that scientific realism about historical sciences is in pretty bad shape (and certainly in worse shape than scientific realism about electrons and genes).

Turner takes the past to be ontologically problematic too. There is a certain sense in which we should take the title of his book entirely literally: scientists do *make* prehistory. To be sure, that's not something Turner affirms. But he does not deny it either. He is neutral on this matter. It is a consistent hypothesis that the past is constructed and, for all we know, it might well be constructed. Turner's meta-philosophical stance is anti-metaphysical. Both realism and anti-realism (in all of its guises) impose metaphysical construals on scientific existential claims. They are not contended with saying that something exists or is real. Realists add a mind-independence gloss on existence/reality and anti-realists render existence/reality mind-dependent. Turner will have none of this. Does that remind you of something? Right guess! He is a fan of the *Natural Historical Attitude*, which is (a) an agnostic attitude towards metaphysical questions and (b) a 'gnostic' stance towards historical knowledge – we do have some knowledge of the past, though its extent should not be exaggerated, as the very idea of an epistemic access to the past is inexorably subject to the aforementioned asymmetries.

As noted already, Turner introduces an epistemic distinction between the past and the microphysical (the tiny). He claims that we can know more about the tiny than the past; hence, it is safer to be a scientific realist about the tiny unobservable. When it comes to the past, the defensible position is what he calls 'historical hypo-realism'. But are past things (e.g., dinosaurs) unobservable? Received wisdom has it that they are not – at least in the sense that they could be observed by suitably placed observers. Turner disputes received wisdom and claims that dinosaurs and their ilk *are* unobservable. Moreover, he argues that there are two distinct types of unobservable – the tiny and the past – and that this typical distinction bears an epistemic weight. Unobservables of type P(ast) are more difficult to be known than of type T(iny).

But is this quite right? Dinosaurs clearly are unlike electrons in terms of unobservability. The sense in which dinosaurs cannot be seen by naked (human) eye is different from the sense in which electrons cannot – different sort of modalities are involved here. Some laws of nature would have to be violated for either of them to be seen, but (interestingly) seeing dinosaurs (but not electrons) would not require a significant violation of the fundamental laws of nature. The possible world in which dinosaurs are observed is closer to the actual than the possible world in which electrons are observed. Observability is a matter of degree, but if we care to make a partition among the actually observed, the observable and the unobservable, dinosaurs are closer to the middle than electrons. In any case, are there epistemically distinct types of unobservables? This cannot be an intrinsic difference of course; it will have to do with a principled difference between how an entity can be known by humans. Hence, it will have to do with the methods used to know that something is the case.

Carol Cleland (2002), based on the Lewisian thesis of the asymmetry of overdetermination, namely, that the present overdetermines the past but it underdetermines the future, has argued quite persuasively that there is, after all, an epistemic *symmetry* between knowing the past and knowing the present (and the future); hence, there is an epistemic symmetry between the methods of historical sciences and the method of physical (experimental, as she would put it) sciences. The idea is that historical scientists explore the present-to-past overdetermination to look for a tie-breaker between rival past hypotheses (a trace entailed by one but not by the other), whereas experimental scientists exploit the present-to-future underdetermination to devise experiments and establish predictions that can tell competing hypotheses apart. On Cleland's view, there is no principled difference between the two methods and both historical sciences and experimental sciences have an equal claim to justified belief and knowledge.

Turner disagrees with all this. What he offers as a reply, however, is not entirely convincing. He thinks (rightly) that there is widespread local underdetermination of past hypotheses. But he goes on to say that this kind of underdetermination is 'less common in experimental science' (Turner 2007, 57). Why? Because in historical sciences, unlike in experimental sciences, scientists cannot manufacture a crucial experiment. Asymmetry number 2 implies that we know that we have irrevocably lost crucial information about the past. There is, indeed, a difference here. But, is it not overstated? First, technological advancements (e.g., computer simulation) can provide plausible missing information about past processes. Second, the manipulation of the tiny can help break underdetermination ties only if we are allowed to bring into the picture the disparate theoretical virtues of competing theories. But we can do exactly that for competing hypotheses about the past.

Turner has interesting responses to this and other objections to his argument so far. For instance, he draws a distinction between a *unifier* (an entity that plays a unifying role) and a *producer* (an entity that can be manipulated to produce new phenomena), and argues that past (un)observables (like dinosaurs) can at best be unifiers whereas tiny unobervables can be producers too. On the basis of this he argues that abductive arguments for past posits will be weaker than abductive arguments for tiny posits. Here is a worry, though. What about the past *and* the tiny, for example, a short-lived lepton? It is posited to explain (by unification, let us say) something that has already happened and though it is manipulable (in principle, at least) it was not manipulated in any way. Doesn't the same hold for a token of the type T-Rex? *Qua* a type of entity a lepton is both a unifier *and* a producer, though some tokens of it are posited as unifiers and others as producers (or both). The same goes for dinosaur (*qua* type): it is both unifier *and* producer, though (it seems that) all tokens of it are posited as unifiers and none as (actual) producers – though they did produce and they are subject to hypothetical manipulation.

Turner presents a number of distinct motivations for adopting forms of 'constructivism' – from Berkeley's, to Kant's, to Kuhn's, to Dummett's, to Latour's. Not all of them are, of course, constructivists and to the extent they can be lumped together as constructivists (anti-realists) their differences might be more significant than their similarities. The bottom-line of Turner's argument is that one may well remain agnostic on the issue of whether the past is real or constructed. This might be surprising for a reason that Turner does not seem to note. The discussion, in the bulk of the book, of local underdetermination, of the information-destroying processes etc. that are supposed to place the historical sciences in an epistemically disadvantageous position requires a sort of realism about the past. What sort? That *there are* historical matters of fact that would make one of the two (or many) competing theories true, but that somehow these facts cannot be accessed. If these facts are not independent in some relatively robust sense (an evidence-transcendent sense, at least), if they are 'socially constructed' as we go along, there is no reason to think that there will be (worse: there *must* be) significant gaps in the past. If facts about the past change over time, or if facts about the past are brought in and out of existence by scientists, it is not obvious to me why the past *resists* its incorporation into theories.

There is something quite puzzling in Turner's treatment of the issue of mind-independence. Here is (roughly) how he sets things up. Realists say: there are Xs (or X occurred), *and* they exist independently of the mental (or X occurred independently of the mental). Constructivists say: There are Xs (or X occurred), *and* they exist in a mind-dependent way (or X occurred mind-dependently). Given this set-up, he complains that the bits that occur after the 'and' are metaphysical add-ons to perfectly sensible scientific claims; they are not empirical hypotheses; they are not confirmable by the evidence etc., etc. But these are not add-ons! Better put: they do not have to be construed as add-ons to perfectly legitimate empirical claims. Turner (2007, 148–9), to his credit, does note that this would be a natural reply, namely, that what is taken to be an add-on is really a way to unpack an existential claims (and there may well be different ways to unpack such claims). But what he says in reply (ibid., 149) seems to miss the point.

Turner's considered claim is that we are faced with a more general case of local underdetermination: scientific theories underdetermine the choice between realism (mind-independence) and social constructivism (mind-dependence). On top of this, we are told, there is reason to think that 'information about whether something happened mind-dependently or mind-independently will never get preserved in the historical record' (ibid., 156). I take it that Turner has some fun here. Too bad that universals, numbers, events and all the rest of the ontic categories do not leave any traces. If Turner is right, all metaphysics is killed off! Perhaps, Turner would be better off if he looked into the logical empiricist tradition of distinguishing between empirical realism and metaphysical realism. One may well be able to leave metaphysics behind without simply being neutral on the realism–constructivism issue.

The Natural Historical Attitude (NHA) that Turner defends is partly an antidote to constructive empiricism. Turner argues (following Kitcher) that constructive empiricism implies scepticism about the past. This, however, depends on whether we think that past posits are observable or not. And we have discussed that already. It is welcoming news that NHA (like its parent, the Natural Ontological Attitude) is not a sceptical stance. The key idea behind this anti-sceptical stance is what Turner calls 'the Principle of Parity' (what I have called the no-double-standards principle), namely, the very same methods of confirmation apply to claims that purport to refer to both observable and unobservable entities and hence that claims about unobservables can be as well supported by the relevant evidence as claims about observables.

## A3. Constructivist anti-realism: a short critique

Christopher Norris's (2004) book fills an important gap in the debate on scientific realism by looking into Norwood Russell Hanson's philosophy of science.[17] Though Norris does not quite put it that way, I think Hanson's key (though neglected before Norris's book) contribution to the debate was that he made possible a non-sceptical version of scientific anti-realism. Put in a nutshell, Hanson's idea is that one can believe in whatever is posited by modern science *and* accept that theories are, by and large, true, while avoiding metaphysical commitments to a mind-independent world and robust realist accounts of truth. The details of Hanson's position are intriguing and subtle and Norris does an excellent job in describing (and criticising) them. But it is useful to keep the big picture in mind, if we are to assess Hanson's contribution. Hanson died prematurely and we can only speculate as to how his views might have developed. But, as Norris amply illustrates, he felt the tensions and problems of the position he was trying to develop.

Hanson was deeply influenced by the later Ludwig Wittgenstein, and in his turn, he deeply influenced Thomas Kuhn and Paul Feyerabend. He employed centrally the Wittgensteinian idea that there is not a ready-made world. Rather, what there is and what one is committed to depends on the 'logical grammar' of the language one uses to speak of the world. Wittgenstein's 'logical grammar' was meant to capture the interconnections of the uses of key concepts that structure a certain language-game. Science is no less a 'language game' than others. This game is characterised by its norms, rules, practices and concepts, but all these are *internal* to the game: they do not give the language-users purchase on an independent world. One can then play the science language-game and adhere to its norms and practices. One can follow the scientific method (whatever that is) and come to accept theories as true as well as believe in the existence of unobservable entities. One, that is, can behave as a scientific realist: one need not be a sceptic. But, on Hanson's view, one need not (perhaps, should not) add to this behaviour any robust realist metaphysics. Nor should one build into the language-game a concept of truth that is evidence-transcendent.

Hanson's philosophy of science has three important entry points. The *first* comes from the rejection of the empiricist view that there can be a theory-free observational language. In fact, Hanson went much beyond this negative thesis. Based on Wittgenstein's claim that all seeing is 'seeing as', he argued that all perception is aspect-relative: there is no way in which Tycho Brahe and Kepler saw the 'same thing' when they turned their eyes to the heavens, since eyes are *blind* and what they see depends on what conceptual input shapes their seeing. This positive thesis leads quickly to claims of incommensurability. In fact, as Norris points out, it leads quickly to perceptual relativism, which renders impossible any attempt to make sense of the empirical basis of science in a way independent of the language-game we adhere to. This first entry point loses the world as a mind-independent structured whole, but re-instates a paradigm-relative world, namely, a world of the phenomena as they are shaped by a certain language-game.

Hanson's *second* entry point comes from quantum mechanics. He bought into the orthodox (Copenhagen) interpretation of it and thought that this leads to inevitable changes in the way we see the world and the way we raise epistemological questions. Presumably, quantum mechanics reveals the inherent limitations of the claim that the world is objective and mind-independent. It also sets limits to what can be known of it and to what kind of theories can be true of it. Norris discusses in some detail Hanson's disapproval of any Bohm-like theory of quantum phenomena and his commitment to a radical discontinuity between the quantum world and the classical one. Here again, Hanson drew the conclusion that accepting quantum mechanics (in its orthodox interpretation) amounts to adhering to a new language-game in light of which the old (classical) language-game makes no sense.

Finally, Hanson's *third* entry point comes from his work on the 'logic of discovery'. In his *Patterns of Discovery* (1958), he did perhaps more than anyone else to legitimise abduction, namely, the mode of reasoning according to which a hypothesis is accepted on the basis that it offers the best explanation of the evidence. Hanson was no friend of instrumentalism. He had no problem with taking scientific theories at face-value. He had no qualms about scientists' going beyond the observable evidence and accepting the existence of unobservable entities on an abductive basis. These unobservable entities are neither logical fictions nor merely hypothetical. They are part of the furniture of the world. But of which world? Hanson's answer is again tied to the idea of language-games: the world as specified by the language-game of science. This world is infested with causal-explanatory connections that underlie the legitimate uses of abductive reasoning, but these connections are, again, the product of several linguistic rules and practices.

It is not hard to see how these three entry points make possible a non-sceptical version of scientific anti-realism: science is not in the business of discovering the structure of a mind-independent world. Rather, it is the language-game that imposes structure onto the world and specifies what facts there are. Accordingly, science can deliver truth, but the truth it does deliver is determined by the epistemic resources, practices and norms of the language-game that constitutes science. This is more evident, as Norris (2004, 113 & 115) notes, in Hanson's notion of a *pattern*. Patterns (the ways objects are conceived) have empirical implications but they are *not* themselves empirical: they are imposed by the conceptual scheme and to deny them is to attack the conceptual scheme itself. There is a Kantian ring in this view. But in Hanson's case, the result is relativised Kantianism. For, in light of perceptual relativism and incommensurability, there is a plurality of language-games each of which constitutes its own phenomenal world. The 'objective' world is either lost or reduced to a *noumenal* blob.

This last point brings to the fore a central problem that Hanson's anti-realism faces: how is change explained? To his credit, Norris (see especially 2004, 37–9) makes capital on this problem on behalf of scientific realists. Here is how I would put the matter. Hanson's view comes to this:

*Constitution*: The worldly objects that science studies are (conceptually) constituted as objects by the language-game (conceptual scheme, rules, theories and practices) that scientific theories use to study the world.

This thesis, however, is in tension with an empirical fact:

*Refutation*: The conceptual schemes that science uses to study the world are revis*able* and revis*ed*.

If scientific objects were constructed/constituted by the conceptual resources of theories, would it not be natural to expect that the very same conceptual resources of the theory would be able to constitute all relevant objects? In particular, how can there be friction *within* the conceptual scheme? Would any friction be either impossible or else explained away by the right constitution of objects? Why, for example, if the relevant scientific objects are constituted by Tycho Brahe's framework, should some phenomena lead scientists to *abandon* this framework? There is a very strong intuition, I think, that the friction can only be explained if the world (something external to the conceptual scheme) exerted some *resistance* to our attempts to conceptualise it.

This intuition, together with *Refutation*, might be thought enough to refute *Constitution*. But there seems to be a way out for its advocates. It might be argued that the world is indeed there, but only as a structure-less (or minimally structured) mould. Yet, this is no improvement. Suppose that the world is a structure-less (or minimally structured) mould. We know that the presence of anomalies to scientific theories is *diachronic*. Anomalies do not go away too easily. Sometimes several modifications of our current theory/conceptual scheme have to be tried before we hit upon the one that removes the anomaly. Besides, anomalies do occur in the theories/conceptual schemes that replace the existing ones. If the world were merely a structure-less mould, then this recurring friction could not be explained. A structure-less mould can be shaped in any way we like. And if it is shaped in a certain way, there is no reason to expect that the shaping will turn out to be inadequate, unless the mould has already, so to speak, a *shape* – a natural causal structure. If the world has a certain causal structure, it is easier to explain why some attempts to fix an anomaly are better than others, as well as to explain why some such attempts prove futile. Hence, if we allow the world to enter the picture as the explanation of friction (and of the subsequent replacement of one's preferred phenomenal world), we'd better also think that this world has already built into it a natural causal structure.

# 2
# Scientific Realism and Metaphysics

There have been two ways to conceive of what scientific realism is about. The first is to see it as a view about scientific theories; the second is to see it as a view about the world. Some philosophers, most typically from Australia, think that the second way is *the* correct way. Scientific realism, they argue, is a metaphysical thesis: it asserts the reality of some types of entity, most typically unobservable entities. I agree that scientific realism has a metaphysical dimension, but I have insisted that it has other dimensions too. In Psillos (1999), and in the previous chapter, scientific realism is characterised thus:

*The Metaphysical Thesis*: The world has a definite and mind-independent structure.

*The Semantic Thesis*: Scientific theories should be taken at face value.

*The Epistemic Thesis*: Mature and predictively successful scientific theories are (approximately) true of the world. The entities posited by them, or, at any rate, entities very similar to those posited, inhabit the world.

This characterisation meshes scientific realism as a view about the world with scientific realism as a view about theories. Taking scientific realism as a view about theories is *not* metaphysically neutral. Yet, scientific realism does not imply *deep* metaphysical commitments. It does not imply commitment to physicalism or to a non-Humean metaphysics.

Ellis (2005) takes my understanding of scientific realism to task. He takes scientific realism to be a view about the world and claims that taking it as a view about theories is wrong. He goes on to argue that scientific realism should be seen as a rich metaphysical world-view that commits realists to physicalism and non-Humeanism.

Though I think Ellis raises important challenges, I disagree with his overall perspective. I accept physicalism and am very sympathetic to

Humeanism. But I do not think that scientific realism should be committed to any of these. These are important issues that should be dealt with independently of the issue of realism in general and of scientific realism in particular. This does not imply that there are no connections between these issues. But it is only when we put these connections into proper perspective that we see what they are.

To put my prime point briefly, the tendency to take scientific realism to be a richer metaphysical view than it is (ought to be) stems from the fact that there are *two* ways in which we can conceive of reality. The first is to conceive of reality as comprising all *facts* and the other is to conceive of it as comprising only *fundamental facts*. I will explain these two senses shortly. But my diagnosis is that scientific realism should be committed to a factualist view of reality and not to a fundamentalist view of it.

## 2.1 A factualist conception of reality

Ellis starts with a well-taken distinction between truth and reality. Truth is attributed to our *representations* of the world. Reality is attributed to the world. Yet, this difference does not foreclose a link between truth and reality. On the *factualist* conception of reality, according to which what is real is what is factual (reality being the totality of facts), there is a two-way traffic between truth and reality. Reality is the realm of facts (truth-makers)[1] and to say that a representation of it is true is to say that it represents a fact: we can go from truth to the facts and from the facts to truth.

This, we might say, is a *metaphysically loaded* conception of truth. Ellis favours a 'metaphysically neutral' conception of truth. He equates truth with epistemic rightness. Though cast in epistemic terms, this conception of truth is *not* metaphysically lightweight. Undoubtedly, judgements about the truth of a matter can (and should) be based on the empirical evidence there is for or against it. But judgements of truth are different from truth. The difference is already there in mundane cases, but it becomes forceful if we consider limiting cases. Suppose we are at the limit of scientific inquiry and claim that all evidence (empirical plus theoretical virtues) for the truth of a theory is in. Suppose we say this theory *is* true. When we reflect on this idealised situation, there are two possibilities. The *first* is that the ideal (epistemically right) theory cannot possibly be false. The *second* is that it is still possible that it be false. If we take truth to be an epistemic concept, it is no longer open to us to think of the second possibility as genuine. This move amounts to a certain metaphysical commitment permeating a seemingly metaphysically neutral conception of truth. The metaphysical character of this commitment becomes evident if we take seriously the second possibility noted above. It amounts a possibility of a *divergence* between what there is in the world and what is issued as existing by an epistemically right theory, which is licensed by the (best) evidence or other epistemic criteria. This possibility captures in the best way the realist claim that truth is answerable to an independent world. The pertinent point is that this is a metaphysical possibility and hence its negation (the first possibility noted above) is also metaphysical.

The possibility of divergence implies an evidence-transcendent understanding of truth. It might be argued that as a matter of fact, whatever is issued by an epistemically right theory is what *really* exists in the world. The realist can accommodate the envisaged convergence by taking the right side in the relevant Euthyphro contrast: Is the world what it is because it is described as thus-and-so by an epistemically right theory *or* is a theory epistemically right because the world is the way it is? At stake here is the order of dependence. Realists should go for the second disjunct. This move makes the world the determinant of epistemic rightness. But this move is not available if the conception of truth is epistemic. My first conclusion is two-fold. On the one hand, the conception of truth that Ellis favours is not metaphysically neutral. On the other hand, it is at odds with some basic commitments that realists should endorse.

I do not think that Ellis's project depends on his epistemic conception of truth. His main argumentative strategy stems from his claim that truth is not metaphysically transparent. To say that the proposition *p* is true is not yet to say *what it is* that makes it true.

## 2.2 A fundamentalist conception of reality

There is a kernel of truth in the claim that truth is not metaphysically transparent. To capture this kernel, let me introduce another way of conceiving reality. We may think of reality as comprising only whatever is *irreducible, basic* or *fundamental*. Accordingly, it is only a sub-class of truths that leads to *the* facts. Reality can still be taken to be the realm of facts, but this is an *elite* set of facts. Under this fundamentalist approach, the factualist conception of reality noted above is not enough to give realism. Some contested propositions are true; hence they represent *some* facts. But what facts they represent is *not* metaphysically transparent; they might represent different (more fundamental) facts than it appears. Truth, in other words, is not metaphysically transparent for the following reason: a true proposition will represent some facts, but it won't necessarily represent them *perspicuously*. The very distinction between representing a fact and representing a fact perspicuously is the kernel of truth in the claim that *truth is not metaphysically transparent*. This line of thought leads to a bifurcation: one might take a reductive or an eliminative attitude towards a set of putative facts.

Let's start with reductivism, as a form of (or a vehicle for) fundamentalism. A reductivist is not, *ipso facto*, an anti-realist. She will be an anti-realist only if she believes that the reduced facts somehow *lose* their factual status. But this is not necessarily so. If reduction is identity or supervenience, the reduced facts do not cease to be facts. On the contrary, far from having their factuality contested, the factuality of the reduced facts is legitimised. If reduction is taken to *remove* factuality, then it amounts to elimination, which is a different story. If elimination is taken seriously it should not be taken to imply that some putative facts are reduced to some other facts. It must be taken to imply that reality is empty of these putative facts. An eliminativist might (and most typically will) grant that there are facts in the world but she will deny that these facts are the truth-makers of the contested propositions. At the same time, an eliminativist will not necessarily deny that the contested propositions purport to refer to facts. She will claim that they *fail* to do so, since there are no relevant facts. That is, she will claim that the contested propositions are *false*. An eliminativist might also find some *use* for these false propositions, but this will not alter the claim that they are false.

If reductivism is distinguished from eliminativism, we can be clear on what reduction achieves: it removes the *sui generis* character of some facts. For instance, there are no *sui generis* mental facts, if the identity theory of mind is true. Similarly, there are no *sui generis* mathematical facts, if logicism is true. But from the claim that '7 + 5 = 12' does not represent a *sui generis* mathematical fact it does *not* follow that it does not represent a fact. Reduction does not show that something is unreal. It shows that it is not *sui generis*. Differently put, reduction shows (or supports the claim) that the contested class of propositions is metaphysically *untransparent*; not that it is *untrue*. So reductivism is *not* anti-factualism.

## 2.3 Factualism vs fundamentalism

The factualist and the fundamentalist conceptions of reality are *not* the same: they are not logically equivalent. One can adopt a factualist view without *ipso facto* being committed to the view that there is an elite class of fundamental facts (or a hierarchical structure of facts). One can be pluralist about facts. There is no logical obstacle in accepting that facts of a more fundamental level are suitably connected with facts of a less fundamental level, without thereby denying the reality of the less fundamental facts. The converse, of course, does not hold. Admitting an elite class of fundamental facts *entails* a factualist view of reality (though restricted to the truths about the elite class). But the difference between the two conceptions of reality suggests that there is need for an independent argument for the claim that facts can be divided into more and less fundamental or for the claim that the only facts there are are the members of the elite class of fundamental facts.

I take it that fundamentalism acts as a *constraint* on one's conception of reality. The primary component of realism is *factualism*. In light of the possibility that a set of propositions may not perspicuously represent the facts, a realist about them must start with an anti-fundamentalist *commitment*. She must take it to be the case that, until further notice, she deals with *not-further-reducible* facts. To put it more linguistically, before one reads off any metaphysical commitments from a true proposition, one must choose to take this proposition at face value. This is a commitment (hence, it can be revoked) to take truth as metaphysically transparent *in the first instance*. Though there is indeed a kernel of truth in the claim that truth is *not* metaphysically transparent, one *can* start with a commitment to its metaphysical transparency, if one starts with a factualist conception of reality and a face value understanding of the propositions employed to represent it. Then, a *conceptually separate* debate can start. If the contested propositions turn out to be metaphysically untransparent, a realist will not cease to be a realist if she argues that their truth-makers are not those implied by a literal reading of these propositions, provided she also holds that these truth-makers *ground* the facts that were taken to be implied by the literal understanding.

If we keep the distinction between factualism and fundamentalism in mind, a number of philosophical benefits follow. *First*, we can put the realism debate in proper focus: realism is about what is real and not about what is fundamentally real. *Second*, we can be clear about the metaphysical commitments that accompany a realist stance about a certain domain: there are genuine facts that make true the propositions of this domain. To say of a fact that it is genuine is to say that it cannot be eliminated from ontology. The right (realist) attitude for a given set of contested propositions is to start with a commitment that it *does* represent genuine facts and then to engage in the independent debate about whether they are *sui generis* or not. If it is shown that these genuine facts are not *sui generis* (if, that is, there are some more fundamental facts that render the contested propositions true), this might revise our deep metaphysical commitments but *not* our claims to truth and reality. *Third*, we can be realists about a number of domains (or subject-matters) without necessarily taking a stance on independent metaphysical issues. The issue of whether some entities are basic, or derivative, or irreducible, or *sui generis*, is a *separate* concern and needs to be addressed separately. It will stem from *other* metaphysical commitments one might have, for example, a commitment to physicalism, or naturalism, or materialism, or pluralism. *Fourth*, there is a clear sense in which one can be an anti-realist about a number of domains (or subject-matters). She will take a contested set of propositions at face value and she will *deny* that there are facts that make them true.[2] But an anti-realist need not be driven by a fundamentalist conception of reality. She need not think that the contested propositions are false because they fail to represent some *fundamental* facts. It is enough that they fail to represent any facts – more specifically, those implied by the literal understanding of them. Of course, someone might start with a fundamentalist conception of reality. But this would lead to anti-realism about a set of putative facts only if some eliminativist stance towards them was adopted.

Hence, though I agree with Ellis that 'the real work has yet to be done', I doubt that this real work falls within the (scientific) realism debate *per se*.

## 2.4 Mind-independence

Realism has been taken to assume that the real is mind-independent. This is partly due to historical reasons. Realism has been taken to be opposed to idealism, the view, roughly put, that whatever exists is mind-dependent because only mental stuff exists. I think idealism is best construed as a kind of fundamentalism and its proper contrast is *materialism*. Berkeley was an immaterialist, after all.

It is not helpful to understand mind-independence in terms of some descriptions that facts should satisfy (or in terms of some characteristic that they may possess). That is, to describe the facts as physical (or material) or as non-mental does not help us understand what it is for them to be mind-independent. In support of this, let us consider the case of modern verificationists. They do *not* doubt that middle-sized objects exist and are irreducibly physical. Yet, they render their reality mind-dependent in a more sophisticated sense: what there is in the world is determined by what can be known (verified, warrantedly asserted) to exist. At stake is a *robust* sense of objectivity, namely, a conception of the world as the arbiter of our changing and evolving conceptualisations of it. It is this sense of objectivity that realism honours with the claim of mind-independence. The world is conceived as comprising the truth-makers of our propositions (allowing, of course, for the possibility that there are truth-makers for which we don't have, and may not have, truth-bearers).

How then should the claim of mind-independence be cast? As noted already in Chapter 1, Section 1.1.3, it should be understood as logical or conceptual independence: what the world is like does not logically or conceptually depend on the epistemic means and the conceptualisations that are used to understand it. This implies, as we have already noted, a commitment to the possibility of a *divergence* between what there is in the world and what is issued as existing by a suitable set of conceptualisations, epistemic practices and conditions. Modern verificationist views foreclose this possibility of divergence by accepting an epistemic conception of truth. Can realists capture the kernel of mind-independence without taking a stand on the issue of truth? I doubt it, for reasons already canvassed by Barry Taylor (1987).

These points have an obvious bearing on Ellis's claim that realism is independent of (a substantive non-epistemic conception of) truth. There is no logical obstacle for a verificationist anti-realist to accept Ellis's physical realism if it is seen as issuing in claims about what exists. To block this possibility, a physical realist should appeal to the mind-independence of these entities (or facts, as I would put it) and, if what said above is right, this is best captured by means of a non-epistemic conception of truth.

Devitt (1997), who, like Ellis, takes realism to be primarily a metaphysical position, has insisted on the claim that the doctrine of realism involves no theory of truth. What has been noted above is that taking realism to involve a non-epistemic conception of truth captures the realist claim of mind-independence. Devitt agrees that realism involves this claim, but argues that its content can be captured without reference to truth. He says that realists can simply deny '*all* dependencies of the physical world on our minds', allowing of course that there are 'certain familiar *causal* relations between our minds and the world' (1997, 306). However, even if it were granted that Devitt's approach avoided the concept of truth in characterising realism about the *physical* world, it cannot characterise the realist stance *in general.* A realist about morality, for instance, might concede that moral principles wouldn't exist if people with minds did not exist. She might claim that there is a sense in which moral principles depend on minds. Yet, she could still be committed to a realist sense of mind-independence if she adopted the foregoing possibility of a divergence between what we (or people, or communities) take (even warrantedly) moral principles to be and what these moral principles are. Casting this possibility of divergence in terms of a non-epistemic conception of truth about moral principles would secure her realism (i.e., the claim that moral principles answer to some moral facts) and, with it, a certain plausible understanding of the claim that moral principles are mind-independent.[3]

Can we at least grant that Devitt's claim avoids the concept of truth in characterising realism about the *physical* world? Devitt's realism implies certain existential commitments, and nothing more. His common-sense realism implies that cats exist, and tables exist etc. His scientific realism implies that electrons exist and quarks exist etc. Though existential assertions *are* ontically committing, there is an ambiguity in claims such as electrons exist. The ambiguity does not concern electrons but *existence.* As noted above, a modern verificationist can (and does) accept that electrons exist. Their gloss on *existence* is that it does not make sense to talk about the existence (or reality) of electrons unless we understand this assertion to mean that..., where the dots are replaced with a suitable epistemic/conceptual condition. Putnam's favourite filling would be based on the condition of rational acceptability; Dummett's would relate to warranted assertibility; and Rescher's would relate to a cognisability-in-principle standard. These views oppose idealism and phenomenalism. They entail (or at least are consistent with the claim) that material objects are real (be they the middle-sized entities of common sense or unobservable entities). The substantive disagreement between them and realism is bound to concern the attribution of existence. In denying anti-realism, it is not enough for Devitt's realism to claim that electrons exist independently of all conditions an anti-realist might specify. There might be an open-ended list of such conditions (with more of those to be specified). What matters to their being *anti-realist* conditions is not that they make existence dependent on something but that they make existence dependent on suitable epistemic/conceptual conditions. It is this *core* of the anti-realist gloss on existence that realists should deny and the best way of doing it is to build into their realism a non-epistemic conception of truth.

Devitt (1997, 54 & 109) acknowledges that existential assertions such as the above commit to the existence of the entities they are about *only if* they are taken at face value. He might well be right in saying that taking them at face value does not imply that we endorse a full and developed semantic theory about them (cf. 1997, 51). It might be enough, as he says, to understand them. This is very close to the realist commitment mentioned above. But note that this commitment is not as innocent as it might seem. As already noted above, it implies that truth is metaphysically transparent *in the first instance*. When I say that electrons exist, I take this to commit me to *electrons* and not, in the first instance at least, to something else. Semantics (and truth) enters the realist position from the front door, by issuing a literal understanding of the existential assertion.

## 2.5 Scientific realism revisited

What it is to be a *scientific* realist? By making a claim to *realism*, scientific realism must make a point about what there is. But how are we to understand the qualifier *scientific*? It refers to science, of course. But can we talk about science in general and what it commits us to? A coarse-grained sense that can be given to scientific realism is to say that it asserts the reality of *unobservable* entities: there are genuine facts that involve unobservable entities and their properties. But note the oddity of describing scientific realism this way. I do not, of course, doubt that there are unobservable entities. But isn't it odd that the basic realist metaphysical commitment is framed in terms of a notion that is epistemic, or worse, pragmatic? This oddity can be explained by reference to the historical development of the scientific realism debate, and more particularly by the fact that some empiricists thought that it is problematic to refer to unobservable entities or that scientific assertions should be rendered *epistemically* transparent by being made to refer to observable facts. These empiricist (mis)conceptions might explain why the scientific realism debate took the turn it did. They do not, however, justify thinking of scientific realism as having to do with the reality of *unobservable* entities.

Note also that, as it stands, the coarse-grained view of scientific realism does not commit a scientific realist to any particular unobservables. It implies only the claim that facts about unobservable entities are among the set of facts. To say something more specific, as I think we should (say about electrons or tectonic plates or genetic inheritance) we need to start with a more determinate view of reality. The issue is not really whether unobservables are real, but rather whether electrons etc. are real. To start from a more *determinate* conception of reality means to start with *specific* scientific theories (or subject-matters, if you like). We should declare our commitment to take them at face value and to make a factualist claim about them (which amounts to arguing that they are true).

All this can be done at two levels of generality. The *less* general level concerns individual subject-matters, say physics or economics or biology. This is where the debate should turn. What is it to be a realist about physics, or biology or economics? I will not answer these questions now. But the points I made above suggest that to be a realist about a subject-matter is to take it a face-value factualist stance towards it. It is then clear how one can be a realist, say, about biology without also being committed to fundamentalism. Biological facts *might* be reducible to physical facts, but (a) this is a separate issue; and (b) it does not entail that there are no biological facts or entities.

The *more* general level concerns scientific theories in general. The question is: what is it to be a realist about scientific theories? Here the only essential difference between the less general questions asked above concerns the *scope* of the question: it is addressed to *any* scientific theory. The realist stance is essentially the same: face-value factualism. Given this level of generality, the realist commitment is to the reality of the entities posited by the theories (whatever those may be).

My own characterisation of scientific realism summarised in the opening paragraphs of this chapter is not far from what I have just called face-value factualism. One of its attractions, I flatter myself in thinking, is that it separates the issue of realism from the issue of fundamentalism. Besides, it transpires that the issue of (un)observability is really spurious when it comes to the metaphysical commitments of realism. What difference does it make to the factual status of claims of modern science that they are about *unobservables*? None whatsoever. The real issue is whether there are facts about electrons, and not whether electrons are unobservables. In a parallel fashion, the real issue is whether there are facts about tracks in cloud chambers and not that these tracks are observable.

Ellis notes that my argument for scientific realism is a two-stage one: from the empirical success of science to the truth of its theories, and then to the reality of the things and processes that these theories appear to describe. He suggests that there is a way to cut out the middle-man (truth) and to offer a direct argument for the metaphysical thesis concerning the reality of the entities posited by science.

Let me start with a *qualified* defence of my argument. I do not agree that I offered a two-stage argument. Instead, I offered *two* arguments. This is because I took seriously another version of anti-realism: sceptical anti-realism. Let's not quarrel about whether van Fraassen's constructive empiricism is really sceptical. The point is that there have been significant arguments challenging the capacity of science to track truth. The argument from the underdetermination of theories by evidence and the pessimistic induction (briefly recapitulated in Chapter 1, Section 1.2) are the primary ones. Oversimplifying, their joint message is that the claim that scientific theories are true is not (never) warranted. Again oversimplifying, my argument from (novel) empirical successes to the truth of theories (in respects relevant to the explanation and prediction of these successes) was meant to block the sceptical onslaught. This is what Ellis considers as the first stage in my two-staged argument. But it is an independent (and distinct) argument. My current concern is not with the success of this argument (though I still think it is successful); it is just that it had a certain distinct aim. I admit that I *might* have taken the sceptical challenge more seriously than it deserves. But then I still think that the realist victory cannot be complete if the sceptical challenge is not met.

There was another (distinct) argument that I offered. In fact, I offered a battery of arguments for the literal reading of scientific theories (against reductive empiricism and instrumentalism) and for the claim that the realist conception of truth should not be epistemic. I will not repeat them here. Suffice it to stress the following. Strictly speaking, what scientific realism needs is the truth of the following conditional: *if* scientific theories are true, *then* the entities posited by them are real. Its antecedent requires a literal understanding of theories and a non-epistemic conception of truth. So, I took it that what realists need to do is defend *literal reading plus non-epistemic truth*. It is then a separate and empirical issue (taken care by my first argument) whether the antecedent of the foregoing conditional is indeed true. If it is, by modus ponens, we can detach its consequent.

## 2.6 How strong is the metaphysics of scientific realism?

The key issue between Ellis and myself is whether there can be an argument for realism that avoids reference to truth. So what should this direct argument for realism look like?

Ellis (2005, 381) is very explicit:

> For the question that needs to be addressed is this: How is the sophisticated, relatively stable, scientific image of the world that is the result of the last two or three centuries of scientific work to be explained? Don't look at it theory by theory, I say, and seek to justify the ontologies of the most successful ones in terms of what these theories are able to predict. Look at the picture as a whole.

What then should we be committed to? Ellis (2005, 382) says:

> The emergence of this scientific image of the world really has only one plausible explanation, viz. that the world is, in reality, structured more or less as it appears to be, and, consequently, that the kinds distinguished in it (the chemical substances, particles, fields, etc.) are, for the most part, natural kinds, and that the causal powers they appear to have are genuine.

I fully agree with the *type* of argument Ellis puts forward. In fact, I think it rests on the only workable criterion of reality. It is the *explanatory criterion*: something is real if its positing plays an indispensable role in the explanation of well-founded phenomena. As I will explain in detail in Chapter 5, I take this to be Sellars's (1963) criterion. Yet, there is a *difference* between the explanatory criterion and Ellis's argument. The explanatory criterion is permissive: it does not dictate the status of the facts that are explanatorily indispensable. Nor is it committed to a hierarchical conception of these facts. The explanatory criterion is at work behind well-known indispensability arguments, which is to say that reality is one thing, fundamentality is another. It is one thing to say that $x$ is real because it meets the explanatory criterion, it is quite another thing to say that $x$ is *sui generis* physical, or abstract or mental. Ellis's argument runs these two things together. Otherwise, Ellis needs an independent argument as to why *all* the entities the scientific image is committed to are physical. If there is such an argument, it is only in tandem with it that the explanatory criterion yields physical realism. Recall that physical realism is the view that the world is basically a physical world. As he says: 'It is a world in which all objects are really physical objects, all events and processes are physical, and in which physical objects can have only physical properties' (Ellis 2005, 375). I happen to believe that this right. But this is not the issue we are discussing. Rather, it is whether this conclusion follows from Ellis's argument. He needs an independent argument for physicalism.

Physicalism (for that's what physical realism amounts to) *can* be argued for independently. But it is, I take it, an important conclusion that it is independent of realism and of scientific realism in particular. Can we get an argument for physical realism from current science? Suppose we can. I see no other way of doing it, apart from taking current science (as a totality) at face value and claim that it is true. The argument would be something like this. Current science posits 'things belonging to an elaborate, strongly interconnected, hierarchical structure of categorically distinct kinds (of chemical substances, particles, fields, etc.), and involved in natural processes which themselves are organised in a natural hierarchy of categorically distinct kinds' (ibid., 382); current science should be taken literally; current science is true; *ergo* these things are real.

In any case, I doubt we can get a *direct* argument for physical realism from current science. Does biology imply that there are no *sui generis* biological facts, or does physics imply that *all* facts are physical? Perhaps, physics does imply that *all* entities are physically constituted. But a fact can still be biological if there are biological properties. I doubt that physics *implies* that there are no *sui generis* biological properties. Or, does physics imply that there are no numbers? Hardly. Ellis claims that the move from truth to reality licenses the view that all sort of things are real (platonic numbers, geometrical points, the theoretical entities of abstract model theories etc.) but argues that 'there is no plausible ontology that would accommodate them' (ibid., 377). Here one can turn the tables on him: doesn't that show that the physicalist ontology is too narrow anyway?

Things become worse if physical realism is taken to include an essentially non-Humean metaphysics. Ellis claims that the physicalism of the 1960s needs to be supplemented in various metaphysically inflationary ways and takes it to be the case that the scientific image posits causal powers, capacities and propensities. Even if it is accepted that the scientific image implies that all things are physically constituted, can it also be taken to imply that their properties are powers, that they have essential natures, that there is real necessity in nature?

The idea that scientific realism must imply some strong metaphysical commitments is fostered (at least partly) by the tendency to associate scientific realism with naturalism. To put naturalism crudely, science is the measure of what is real. If naturalism is taken in its extreme form (physicalism), the implication is that only the physical is real. The first thing to be said here is that scientific realism is independent of naturalism. One can be a scientific realism and accept *sui generis* non-physical entities. The second thing to be said is that, though independent, scientific realism and naturalism are good bed fellows. So the important issue is whether naturalism dictates any strong metaphysical views. If you are a naturalist you should take current physics and biology seriously. But does current science imply any commitments to essentialism, dispositions, universals, natural necessity and the like? To put the question differently: does science imply a non-Humean view of the world?

Note, first, an irony. Answering *this* question requires taking science at face value. It requires that science implies that there is necessity in nature, that there are causal powers, essential properties and the like. Even if this were granted, it would still remain open whether these were *sui generis* entities. As noted above, this is an independent issue. A scientific realist can accept, say, causal powers, but argue (separately and independently) that they are reducible to categorical properties of the objects that possess them.

There is not an inconsistency in believing in electrons and in Humean laws and in all powers requiring categorical bases. But it may be thought that scientific realism (or naturalism) is best viewed in tandem with a non-Humean metaphysics. For the time being, I want to remain neutral on this. I don't think science *implies* a non-Humean conception of the deep metaphysical structure of the world. If, for instance, we take the Mill–Ramsey–Lewis view of laws as offering an *objective* and *robust* view of laws (see my 2002, 292–3), then one can be a scientific realist, accept that there are contingent laws of nature and take the Mill–Ramsey–Lewis view of them. Or, if one is a scientific realist, one *must* accept the existence of natural properties, but may take these to be Lewisian natural classes. Or, if one is a scientific realist, one may accept that some properties are powers, but deny that they are ungrounded powers.[4] Or, if one is a scientific realist, one should take causation seriously but think that, ultimately, it is a relation of probabilistic dependence among event-types. Despite all this, I think a scientific realist can be open-minded in the sense that there may well be independent reasons to take a stronger (non-Humean) view of laws, properties, necessity, causation, powers and the like.

# 3
# Thinking About the Ultimate Argument for Realism

The title of this chapter alludes to Musgrave's piece 'The Ultimate Argument for Realism', though the expression is van Fraassen's (1980, 39), and the argument is Putnam's (1975, 73): realism 'is the only philosophy of science that does not make the success of science a miracle'. Hence, the code name 'no-miracles' argument (henceforth, NMA). The NMA has quite a history and a variety of formulations. I have documented all this in my work (Psillos 1999, Chapter 4). No matter how exactly the argument is presented, its thrust is that the success of scientific theories lends credence to the following two theses: (a) scientific theories should be interpreted realistically and (b) so interpreted, these theories are approximately true. The original authors of the argument, however, did not put an extra stress on *novel* predictions, which, as Musgrave (1988) makes plain, is the litmus test for the ability of any approach to science to explain the success of science.

Here is why reference to novel predictions is crucial. Realistically understood, theories entail too many novel claims, most of them about unobservables (e.g., that there are electrons, that light bends near massive bodies, etc.). It is no surprise that some of the novel theoretical facts a theory predicts may give rise to novel *observable* phenomena, or may reveal hitherto unforeseen connections between known phenomena. Indeed, it would be surprising if the causal powers of the entities posited by scientific theories were exhausted in the generation of the already *known* empirical phenomena that led to the introduction of the theory. On a realist understanding of theories, novel predictions and genuine empirical success is to be expected (given of course that the world co-operates).

The aim of this chapter is to rebut two major criticisms of NMA. The first comes from Musgrave (1988). The second comes from Howson (2000). Interestingly, these criticisms are the mirror image of each other. They both point to the conclusion that NMA is *fallacious*. Musgrave's misgiving against NMA is that if it is seen as an *inference to the best explanation* (IBE), it is deductively fallacious. Being a deductivist, he rectifies it by turning it into a *valid deductive argument*. Howson's misgiving against NMA is that if it is seen as an *IBE*, it is *inductively* fallacious. Being a subjective Bayesian, he rectifies it by turning it into a *sound subjective Bayesian argument*. I will argue that both criticisms are unwarranted.

There would be no problem with Musgrave's version of NMA if deductivism were correct. But the deductivist stance is both descriptively and normatively wrong. To avoid a possible misunderstanding, I should note that I have no problem with deductive logic (how could I?). My problem is with deductivism, that is the view that, as Musgrave (1999b, 395) puts it, 'the only valid arguments are deductively valid arguments, and that deductive logic is the only logic that we have or need'. One could cite Bayesianism as a live example of why deductivism is wrong. But, I think, there are important problems with Bayesianism too.[1] Briefly stated, the Bayesian critique of NMA is that it commits the base-rate fallacy. Howson tries to rectify this by arguing that a 'sounder' version of NMA should rely explicitly on *subjective* prior probabilities. Against the Bayesian critique of NMA I will primarily argue that we should resist the temptation to cast the NMA in a subjective Bayesian form. However, I will also explore the possibility of accepting a more objective account of prior probabilities, if one is bent on casting NMA in a Bayesian form.

## 3.1 The no-miracles argument

The structure of NMA and its role in the realism debate is quite complex. To a good approximation, it should be seen as a grand IBE. The way I read it, NMA is a philosophical argument that aims to defend the reliability of scientific methodology in producing approximately true theories and hypotheses. I don't want to repeat here the exact formulation of the argument (see Psillos 1999, 78–81). However, I want to emphasise that its conclusion has two parts. The *first* part is that we should accept as (relevant approximately) true the theories that are implicated in the (best) explanation of the *instrumental* reliability of first-order scientific methodology. The *second* part is that since, typically, these theories have been arrived at by means of IBE, IBE is reliable. Both parts are necessary for my version of NMA.

The main strength of NMA rests on the way the first part of its conclusion is reached. Following more concrete types of explanatory reasoning that occur all the time in science, it suggests that it is reasonable to accept certain theories as approximately true, at least in the respects relevant to their theory-led successes. So, it is successful instances of explanatory reasoning in science that provide the *basis* for the grand abductive argument. However, NMA is not *just* a generalisation over the scientists' abductive inferences. Although itself an *instance* of the method that scientists employ, it aims at a much broader target: to defend the thesis that IBE (i.e., a *type* of inferential method) is reliable. This relates to the second part of its conclusion. What makes NMA distinctive as an argument for realism is that it defends the achievability of theoretical truth. The second part of the conclusion is meant to secure this. The background scientific theories, which are deemed approximately true by the first part of the conclusion, have themselves been arrived at by abductive reasoning. Hence, it is reasonable to believe that abductive reasoning is reliable: it tends to generate approximately true theories. This is not meant to state an a priori truth. The reliability of abductive reasoning is an empirical claim, and if true, it is contingently so.

As I conceive of it, NMA needs a qualification. Although most realists would acknowledge that there is an explanatory connection between a theory's being empirically successful and its being, in some respects, right about the world, it is far too optimistic – if at all defensible – to claim that *everything* that the theory asserts about the world is thereby vindicated. Realists should *refine* the explanatory connection between empirical and predictive success, on the one hand, and truthlikeness, on the other. They should assert that these successes are best explained by the fact that the theories which enjoyed them have had *truthlike theoretical constituents* (i.e., truthlike descriptions of causal mechanisms, entities and laws). The theoretical constituents whose truthlikeness can best explain empirical successes are precisely those that are essentially and ineliminably involved in the generation of predictions and the design of methodology which brought these predictions about. From the fact that not every theoretical constituent of a successful theory does and should get credit from the successes of the theory, it certainly does *not* follow that none do (or should) get some credit.[2]

There are a number of objections to this explanationist version of NMA. One of them has also been pressed by Musgrave (1988, 249; 1999a, 289–90). The objection is that NMA is *viciously* circular: it employs a second-order IBE in defence of the reliability of first-order IBEs. In Psillos (1999, 81–90), I argued that (a) there is a difference between premise-circularity and rule-circularity (a premise-circular argument employs its conclusion as one of its premises; a rule-circular argument conforms to the *rule* which is vindicated in its conclusion); (b) rule-circularity is *not* vicious; and (c) the circularity involved in the defence of basic rules of inference is rule-circularity. The employment of IBE in an abductive defence of the reliability of IBE is *not* viciously circular. As a support of all this consider the following case. Many (if not all) use modus ponens unreflectively as an inferential rule and yet the establishment of the *soundness* of modus ponens proceeds with an argument that effectively uses modus ponens. This procedure can still explain to users of modus ponens why and in virtue of what features deductive reasoning is sound.

Being a deductivist, Musgrave thinks that the *only* kind of validity is deductive validity. He denies that there are such things as cogent non-deductive arguments (cf. 1999b). He takes it that rule-circular arguments in favour of inferential rules may have only some *psychological* force (cf. 1999, 289–90). But Musgrave (1999, 295) is aware of the point that the proof of the soundness of modus ponens requires the use of modus ponens. How does he react to this? He has wavered between two thoughts. The *first* is that 'there is little future in the project of "justifying deduction"' (ibid., 296). As he acknowledges, 'Any "justification" which is non-psychologistic will itself be a deductive argument of some kind, whose premises will be more problematic than the conclusion they are meant to justify' (ibid.) To be sure, he immediately adds that there is a difference between deductive rules and non-deductive (ampliative) ones: even if neither of them can be 'justified', non-deductive rules can be *criticised*. How much pause should this give us? Let us grant, as we should, that none of our basic inferential rules (both deductive and non-deductive) can be 'justified' without rule-circular arguments. The fact that the non-deductive rules can be criticised more severely than the deductive ones may make us be much more cautious when we employ the former. That's all there is to it. The *second* thought that Musgrave has (cf. 1980, 93–5; 1999a, 96–7) is that there is a sense in which deduction *can* be 'justified', but this requires an appeal to 'deductive intuitions'. As Musgrave (1980, 95) graphically puts it: 'In learning logic we pull ourselves up by our bootstraps, exploit the intuitive logical knowledge we already possess. Somebody who lacks bootstraps ("deductive intuition") cannot get off the ground'. This, I think, is exactly right. But, as I have argued in some detail in Psillos (1999, 87–9), exactly the same response can be given to calls for 'justifying' non-deductive rules. When it comes to issues concerning the vindication of IBE, if one lacks 'abductive' intuitions, one lacks the necessary bootstraps to pull oneself up.

## 3.2 Deductivism

Musgrave (1988, 237; 1999a, 60) takes NMA to be an IBE. As noted already, Musgrave (1988, 232; 1999a, 119) has been one of the first to stress that what needs to be explained is *novel* success (i.e., the ability of theories to yield successful novel predictions). And he has been one of the first to note that NMA should focus on the novel success of *particular* theories (cf. ibid., 249). He has also produced some powerful arguments to the effect that non-realists explanations of the success of science are less satisfactory than the realist one. Most of them appear in Musgrave (1988). In fact, he (ibid., 249) concludes that the realist explanation is the *best*. The issue then is this: does Musgrave endorse NMA? The answer to this question is *not* straightforward.

Precisely because Musgrave takes NMA to be an IBE, he takes it to be deductively invalid; hence fallacious. Being a deductivist, he takes it that the only arguments worth their salt are deductive. Hence he cannot endorse NMA, at least as it stands. Musgrave takes all *prima facie* non-deductive arguments to be *enthymemes*. An enthymematic argument is an argument with a missing or suppressed premise. After the premise is supplied (or made explicit), the argument becomes deductively valid, though it may or may not be sound (cf. Musgrave 1999a, 87 & 281ff). According to Musgrave, non-deductive arguments are really deductive enthymemes, with 'inductive principles' as their missing premises.

As it is typically presented, IBE has the following form (cf. Musgrave 1988, 239; 1999a, 285):

*IBE*

(i) F is the fact to be explained.
(ii) Hypothesis *H* explains *F*.
(iii) Hypothesis *H* satisfactorily explains *F*.
(iv) No available competing hypothesis explains *F* as well as *H* does.
(v) Therefore, it is reasonable to accept *H* as true.

Given that this argument-pattern is invalid, Musgrave proposes that it should be taken to be enthymematic. The missing premise is the following epistemic principle (cf. ibid.):

(*missing premise*)

> It is reasonable to accept a satisfactory explanation of any fact, which is also the best explanation of that fact, as true.

Add to (*IBE*) the missing premise, and you get a valid argument. Succinctly put, the deductive version of IBE is this:

(*D-IBE*)

If hypothesis *H* is the best explanation of the fact to be explained,[3] then it is reasonable to accept *H* as true.

*H* is the best explanation of the evidence.

Therefore, it is reasonable to accept *H* as true.

This is a valid argument. Besides, Musgrave (1999a, 285) thinks that 'instances of the scheme might be sound as well'. He takes it that the missing premise 'is an epistemic principle which is not obviously absurd' (ibid.) In light of this, it's no surprise that Musgrave reconstructs NMA as an enthymeme. That's how Musgrave (1988, 239) puts it:

> The fact to be explained is the (novel) predictive success of science. And the claim is that realism (more precisely, the conjecture that the realist aim for science has actually been achieved) *explains* this fact, explains it *satisfactorily*, and explains it *better* than any non-realist philosophy of science. And the conclusion is that it is reasonable to accept scientific realism (more precisely, the conjecture that the realist aim for science has actually been achieved) as true.

This is a deductive enthymeme, whose suppressed premise is the aforementioned epistemic principle (*missing premise*). Musgrave takes NMA to aim to tell in favour of the *Epistemic Thesis* of scientific realism (see Chapter 1, Section 1.1). Though he formulates the argument in terms of his own axiological thesis, he takes it that, if successful, NMA makes it reasonable to accept that truth has been achieved.

I would have no problem with (*D-IBE*) if deductivism were correct. But the deductivist stance is so radically at odds with the practice of science (as well as of everyday life) that it would have to give pause to even the most dedicated deductivist. Human reasoning is much broader than deductivists allow. It is *defeasible*, while deductive reasoning is not. It is sensitive to new information, evidence and reasons in ways that are not captured by deductive arguments. The latter are monotonic: when further premises are added to a valid deductive argument, the original conclusion still follows. Human reasoning is non-monotonic: when new information, evidence and reasons are added as premises to a non-deductive argument, the warrant there was for the original conclusion may be removed (or enhanced). Human reasoning is also *ampliative*, while deductive reasoning is not. The conclusions we adopt, given certain premises, have excess content over the premises. Deductive reasoning is not content-increasing. In a (logical) sense, the conclusion of a valid deductive argument is already 'contained' in its premises.[4] This is not to belittle deductive reasoning. It's the only kind of reasoning that is truth-preserving. The importance of truth-preservation can hardly be exaggerated. But we should not forget that, though deductive reasoning preserves truth, it cannot establish truth. In particular, it cannot establish the truth of the premises. If we are not talking about logical (and mathematical and analytical truths), the premises of deductive arguments will be synthetic propositions, whose own truth can be asserted, if at all, on the basis of ampliative and non-deductive reasoning. Though deductive reasoning is indispensable, it can hardly exhaust the content and scope of human (and scientific) reasoning.[5] As a *descriptive* thesis, deductivism is simply false.

Is then deductivism to be construed as a *normative* thesis? I am aware of no argument to the effect that deductivism is normatively correct. This is not to imply that deductive logic has no normative force. But recall that deductivism is the thesis that *all* arguments worth their salt *should* be construed as deductive enthymemes. Whence could this thesis derive its supposed normative force? I don't see a straightforward answer to this question. Musgrave (1999a, 284–5) suggests that reconstructing supposed non-deductive arguments as deductive enthymemes 'conduces to clarity', by making their premises explicit. Hence, it also makes explicit what is required for the premises to be true, and for the argument to be sound. This point seems problematic. Non-deductive arguments (e.g., simple enumerative induction, or IBE) are *not* unclear. If anything, the problem with them is how to justify them. But a similar problem occurs with deduction, as we saw at the end of the previous section. Suppose that we leave this problem to one side. Suppose we grant that turning a non-deductive argument into a deductively valid one conduces to clarity since it makes its premises explicit. Deductivists still face a problem: what, if anything, justifies the missing premise? To fix our ideas, consider the major premise of (*D-IBE*) above. What justifies the principle 'If hypothesis *H* is the best explanation of the fact to be explained, then it is reasonable to accept *H* as true'? The sceptic can always object to *this* principle that it is question-begging. Perhaps, deductivism is a fall-back position. It asserts that arguments can be *reconstructed* as deductively valid arguments. But this thesis is trivial. *Any* argument can be turned into a deductively valid one by adding suitable premises. In particular, any *invalid* argument can be rendered valid by adding suitable premises. Consider the fallacy of affirming the consequent. The argument:

If (if a and b) and b, then a
If a then b
b
Therefore, a

is perfectly valid. If all logically invalid arguments were considered enthymemes, there would be no such thing as invalidity. Musgrave is aware of this objection, too. His reply is this: '(Y)ou cannot allow anything whatever to count as a "missing premise"; what the "missing premise" is must be clear from the context of the production of the argument in question' (Musgrave 1999b, 399; 1999a, 87, n106). But, surely, the context underdetermines the possible 'missing premises'. More importantly, for any 'missing premise', there will be *some* contexts in which it is appropriate.

To recapitulate: Musgrave's misgivings against NMA were motivated by the thought that if it is seen as an IBE, it is deductively fallacious. He tried to correct it, by turning it into a *valid deductive argument*. We found his attempt wanting because we found deductivism wrong. What is interesting is that others, most notably Howson, think that if it is seen as an IBE, NMA is *inductively* fallacious. He tries to correct it, by turning it into a sound subjective Bayesian argument. All this will leave Musgrave totally unmoved, since he thinks there is no such thing as inductive logic (cf. 1999b). Still, for those of us who (a) think there is more to reasoning than deduction, (b) are critical of subjective Bayesianism and (c) want to defend some form of NMA, it will be important to examine whether the Bayesian criticism of NMA succeeds or fails.

## 3.3 Subjective Bayesianism to the rescue?

Howson (2000, 36) formulates NMA as follows:

(A)

(i) If a theory T is not substantially true then its predictive success can only be accidental: a chance occurrence.
(ii) A chance agreement with the facts predicted by T is very improbable – of the order of a miracle.
(iii) Since this small chance is so extraordinarily unlikely, the hypothesis that the predictive success of T is accidental should be rejected (especially in light of the fact that there is an alternative explanation – namely, that T is true – which accounts better for the predictive success).
(iv) Therefore, T is substantially true.[6]

He then argues in some detail that (A) is inductively fallacious. He contests the soundness of all if its premises (cf. ibid., 43). However, the novelty of Howson's view relates to his criticism of premise (iii) and of the inferential move to (iv). His prime point is that (A) is wrong because it commits the base-rate fallacy.

The base-rate fallacy is best introduced with a standard example in the literature, known as the Harvard Medical School test.

(*Harvard Medical School test*)
A test for the presence of a disease has two outcomes, 'positive' and 'negative' (call them + and −). Let a subject (Joan) take the test and let H be the hypothesis that Joan has the disease and −H the hypothesis that Joan doesn't have the disease. The test is highly reliable: it has zero *false negative* rate. That is, the likelihood that the subject tested negative given that she does have the disease is zero (i.e., prob(−/H) = 0). Consequently, the *true positive* rate, that is, the likelihood of being tested positive given that she has the disease is unity (prob(+/H) = 1). The test also has a very small *false positive* rate: the likelihood that Joan is tested positive though she doesn't have the disease is, say, 5 per cent (prob(+/−H) = .05). Joan tests positive. What is the probability that Joan has the disease given that she tested positive? That is, what is the posterior probability prob(H/+)?

When this problem was posed to experimental subjects, they tended, with overwhelming majority, to answer that the probability that Joan has the disease given that she tested positive was very high – very close to 95 per cent.

This answer is wrong. Given only information about the likelihoods prob(+/H) and prob(+/−H), the question above – what is the posterior probability prob(H/+)? – is indeterminate. This is so because there is some crucial information *missing*: we are not given the incidence rate (base-rate) of the disease in the population. If this incidence rate is very low, for example, if only 1 person in 1,000 has the disease, it is very *unlikely* that Joan has the disease even though she tested positive: prob(H/+) would be less than .02.[7] For prob(H/+) to be high, it must be the case that prob(H) be not too small. But if prob(H) is low, it can dominate over a high likelihood of true positives and lead to a very low posterior probability prob(H/+). The lesson that many have drawn from cases such as this is that it is a fallacy to ignore the base-rates because it yields wrong results in probabilistic reasoning. The so-called base-rate fallacy is that experimental subjects who are given problems such as the above tend to neglect base-rate information, *even when* they are given this information explicitly.[8]

With this in mind, let us take a look at NMA. To simplify matters, let *S* stands for predictive success and *T* for a theory. According to (A) above, the thrust of NMA is the comparison of *two* likelihoods, namely, prob(S/−T) and prob(S/T). The following argument captures the essence of Howson's formulation of NMA (see (A) above).

(B)
prob(S/T) is high.
prob(S/−T) is very low.
S is the case.
Therefore, prob(T/S) is high.[9]

Let us say that the false positive rate is low and the false negative rate is naught. That is, the probability of T being successful given that it is false is very small (say, prob(S/−T) = .05)) and the probability of T being unsuccessful given that it is true is zero (i.e., prob(−S/T) = 0). Hence, the true positive rate (prob(S/T)) is 1. Does it follow that prob(T/S) is high? NMA is portrayed to answer affirmatively. But if so, it is fallacious: it has neglected the base-rate of truth (i.e., prob(T)). Without this information, it is impossible to estimate correctly the required posterior probability. If the base-rate of true theories is low, then prob(T/S) will be very low too. Assuming that base-rate of true theories is 1 in 100 (i.e., prob(T) =.01), prob(T/S) =.17. The conclusion seems irresistible: as it stands, (B) commits the base-rate fallacy – it has neglected prob(T), or as the jargon goes, the base-rate.

Every cloud has a silver lining, however. So, Howson (2000, 55–9) urges us to think how NMA could become 'sounder' within a Bayesian framework. We are invited to accept that NMA can succeed only if information about base-rates (or prior probabilities) is taken into account. In effect, the idea is this:

($B^1$)
prob(S/T) is high.
prob(S/−T) is very low.
S is the case.
prob(T) is not very low.
Therefore, prob(T/S) is high.

What has been added is an explicit premise that refers to the *prior probability* of true theories. For ($B^1$) to be sound, this probability should not be low. How low prob(T) can be will vary with the values of prob(S/T) and prob(S/−T). But it is noteworthy that, with the values of the likelihoods as above, if prob(T) is only 5 per cent, prob(T/S) is over 50 per cent. To be sure, ($B^1$) is not valid. But, as Howson (2000, 57) notes, it is 'a sound probabilistic argument'. ($B^1$) rests also on the assumption that prob(S/−T) is very low. This can be contested. But, Howson notes, there may be occasions on which this low probability can be justified, e.g., when, for instance, we think of −T as a disjunction of *n* theories $T_i$ ($i=1,\ldots, n$) whose own prior probabilities prob($T_i$) are negligible. In any case, his point is that NMA can be a sound argument only when we see that it is based on some substantive assumptions about *prior probabilities*. Being a subjective Bayesian, he takes these prior probabilities to be 'necessarily subjective and a priori' (ibid., 55).

## 3.4 A whiff of objectivism

I will start my criticism of Howson's argument by resisting the view that one needs to rely on *subjective* prior probabilities in formulating NMA. So for the time being at least, I will assume the foregoing Bayesian reformulation of NMA. Let us reformulate ($B^1$), based on what has been called the *Bayes factor*. This is the ratio

(*Bayes factor*)
$f$=prob(S/−T)/prob(S/T).

Recall Bayes's theorem:
prob(T/S) = prob(S/T)prob(T)/prob(S), (1)
where
prob(S) = prob(S/T)prob(T)+prob(S/−T)prob(−T).
Using this factor, (1) becomes this:
prob(T/S) = prob(T)/ prob(T) + $f$ prob(−T). (2)
($B^1$) can then be written thus:

($B^2$)
$f$ is very small.
S is the case.
prob(T) is not very low.
Therefore, prob(T/S) is high.

The Bayes factor is small if prob(S/−T) $\ll$ prob(S/T). Whether or not the conclusion follows from the premises depends on the prior probability prob(T). So, the Bayes factor, on its own, tells us little. But it *does* tell us something of interest; something that can take out *some* to the sting of subjectivism in Bayesianism. Two things are relevant here. The *first* is that there is a case in which the prior probability of a theory does not matter. This is when the Bayes factor is *zero*. Then, no matter what the prior prob(T) is, the posterior probability prob(T/S) is one. The Bayes factor is zero if prob(S/−T) is zero. This happens when just one theory can explain the evidence. Then, we can dispense with the priors. This situation may be unlikely. But it is not a priori impossible. After all, the claim that evidence underdetermines the theory is not a logical truth! Put in a different way, one quick problem that Howson's reconstructions of NMA faces is that it equates, at least implicitly, explanation with deduction. Given this equation, it is trivially true that there cannot be just one theory that explains the evidence, since there will be many (an infinite number of?) theories that entail it. In many places, Howson (cf., for instance, 2000, 40–1) does make this equation. But this is a Phyrric victory over NMA. There is more to explanation than the deduction of (descriptions of) the phenomena from the theory (and deduction is not even necessary for explanation). So, it may well be the case that many theories entail (descriptions of) the relevant phenomena, while only one of them *explains* them. I won't argue for this claim now. Suffice it for the present purposes to note that equating explanation with deduction is question-begging (some details are offered in Chapter 10).

Let us grant that the case in which the Bayes factor is zero is exceptional. There is a *second* thing in relation to Bayes factor that needs to be noted. Assume some kind of indifference (or a flat probability distribution) between prob(T) and prob(–T); that is, assume that prob(T) = prob(–T) = 1/2. Then (2) above becomes:

prob(T/S) = 1/ 1 + *f*. (3)

Assuming indifference, the Bayes factor shows that likelihood considerations (especially the fact, if it is fact, that *f* is close to zero) can make T much more likely to be true. The point here is not that we can altogether dispense with the priors. Rather, it is that we are not compelled to take a *subjective* view of the prior probabilities. So, there is a version of NMA which, though close to ($B^2$) above, does *not* assume anything other than indifference as to the prior probability of T being true.

($B^3$)
*f* is close to zero.
S is the case.
prob(T) = prob(–T) = 1/2.
Therefore, prob(T/S) is high.

If one wanted to capture the *thrust* of NMA within a Bayesian framework, one could hold onto ($B^3$). This does not commit the base-rate fallacy. Besides, it avoids the excesses of subjective Bayesianism.

I have assumed so far that prior probabilities and base-rates are one and the same thing. Howson does assume this too. Howson (2000, 57, n5) calls the prior probabilities 'the epistemic analogue of the base-rate'. Normally, base-rates are given by reliable statistics. Hence, they are quite objective. When a subject is asked how probable it is that Jim (a young adult male) suffers from hypothyroidism, given that he has the symptoms, she doesn't commit a fallacy if she ignores her own prior degree of belief that Jim has hypothyroidism. After all, she might not have any prior degree of belief in this matter. The fallacy consists in her claiming that the probability is high while ignoring some relevant factual information about hypothyroidism, namely, that it is quite rare, even among people who have the relevant symptoms. This is some objective statistical information, for example, that only 1 in 1,000 young adult males suffers from hypothyroidism. Base-rates of this form can (and should) be the input of a prior probability distribution. But they are *not* the prior subjective degrees of belief that Bayesians are fond of. In incorporating them, Bayesians move away from a purely subjective account of prior probabilities. What about the converse? If prior probabilities are purely (and necessarily, as Howson says) subjective, why should an agent rely on base-rates to fix her prior probabilities?

Why should an agent's subjective prior probability of an event to occur be equated with the rate of the occurrence of this event in a certain population? Purely subjective priors might be assigned in many ways (and, presumably, there is no fact of the matter as to which way is the correct, or rational, one). An agent might know a relevant-base rate but, being a purely subjective Bayesian, she might decide to disregard it. She won't be probabilistically incoherent, if she makes suitable adjustments elsewhere in her belief corpus. Or, though the base-rate of hypothyroidism in the population is very low, her subjective prior probability that Jim suffers from hypothyroidism may be quite high, given that she believes that Jim has a family history of hypothyroidism. The point here is that *if* prior probabilities are purely subjective, it seems within the rights of a Bayesian agent to fix her prior probabilities in a way different from the relevant base-rates. So, prior probabilities are not, necessarily, base-rates. Or, more provocatively, ba(y)se rates are *not* base-rates.

In light of this, something stronger can be maintained. Subjective Bayesians had better have a more objective account of prior probabilities, if they are to reason correctly (according to their own standards) and avoid falling victims of the base-rate fallacy. For if prior probabilities are totally up to the agent to specify, the agent seems entitled to *neglect* the base-rate information, or to adopt a prior probability which is significantly lower or higher than the base-rate. If anything, base-rates should act as an external constraint on Bayesian reasoning, by way of fixing the *right* prior probabilities.

## 3.5 Ignoring base-rates

The Bayesian critique of NMA (see argument (B) above) consists in the claim that it *ignores* the base-rates of truth and falsity. But there is a sense in which this is not quite correct. The Bayesian criticism presupposes that *there are* base-rates for truth and falsity. However, it is hard, if not outright impossible, to get the relevant base-rates. The issue is not *really* statistical. It's not that we don't have a list of true and false theories at our disposal. Nor, of course, is the issue that the advocates of NMA fail to take account of such a list. The issue is philosophical. The very idea of a base-rate of truth and falsity depends on how the *relevant* population of theories is fixed. This is where many philosophical problems loom large. For one, we don't know how exactly we should individuate and count theories. For another, we don't even have, strictly speaking, outright true and false theories. But suppose that we leave all this to one side. A more intractable problem concerns the concept of success. What is it for a theory to be successful? There is no reason here to repeat well-known points (see Psillos 1999, 104–8). But the general idea is clear. By choosing a loose notion of success, the size of the relevant population might increase and a lot of false theories might creep in. True theories won't be left out, but they may be vastly outnumbered by false ones. There will be many more false positives than otherwise. In this population, the probability of a randomly selected theory being true will be low. By choosing a stricter notion of success, for example, by focusing on *novel* predictions, fewer theories will be admitted into the relevant population. The number of true theories will exceed the number of false theories. The number of false positives will be low, too. In that population, the probability of a randomly selected theory being true will be high. In sum, base-rates are unavailable not because we don't have enough statistics, but because we don't have clear and unambiguous reference classes. And we don't have the latter because our central individuating concepts (theory, success, etc.) are not precise enough.[10]

Suppose that there are base-rates available. Is it always a bad idea to ignore them? To address this question, let us go back to the original setting of the base-rate fallacy and take a look at another standard case in which this fallacy is to be committed. This is the Blue Cab/Green Cab case.

(*Blue Cab/Green Cab*)
There is a city in which there are two cab companies, the Green cabs and the Blue cabs. Of the total number of cabs in the city, 85 per cent are green and 15 per cent are blue. There was a late-night hit-and-run car accident and the sole eyewitness said that it was a *blue* cab involved. The eyewitness is very reliable: in test situations involving blue and green objects at night, he made the correct identifications in 80 per cent of the cases and he was mistaken in 20 per cent of cases. What is the probability that the culprit was a blue cab?

When asked the foregoing question, subjects involved in psychological experiments, tended to trust the eyewitness and said, in an overwhelming percentage, that the probability that the culprit was a *blue* cab was very high. This is supposed to be a standard case of the base-rate fallacy, since, given the base-rates for blue and green cabs, the probability that the culprit was a blue cab is low (.41). It's more likely that the culprit was a green cab, since there are many more of those around.

There are two points that need to be noted. *First*, it is one thing to reason correctly probabilistically (the subjects, obviously, didn't). It is quite another thing to get at the truth. For, it may well be that the eyewitness really *saw* a blue cab and that a blue cab was involved in the accident. What is important here is that the base-rate information might have to be ignored, if what we want to get at is the *truth*. There is not, of course, any definite answer to the question: when are the base-rates to be ignored and when are not? But there is an interesting observation to be made. In the case at hand, there is some crucial information to be taken into account, namely, that the situation is *ambiguous*. After all, it was dark and, in the dark, our observations are not very reliable. Actually, as Birnbaum (1983) has noted, if a witness is aware that there are many more green cabs than blue cabs in the city, he is predisposed to see green cabs in ambiguous situations. This, it should be noted, is a piece of information (or background knowledge) that the subjects of the experiment also have. The very fact that, despite the prevailing disposition, the witness is reported to have seen a *blue* cab carries *more* weight than the relevant base-rates. There is a sense in which the subjects commit a fallacy (since they are asked to reason probabilistically but fail to take account of the base-rates), but there is another sense in which they reason correctly because the salient features of the *case history* can get them closer to the truth.

Transpose all this to the problem of truth and success. If we take the base-rates into account, we may get at the correct probability of a theory's (chosen at *random*) being approximately true, given that it is successful. And this probability may be quite low, if the base-rate of truth is very low. Suppose we conclude from this that this theory is *not* approximately true (because it is very unlikely that it is). But it may well *be* approximately true. The fact that it appears unlikely to be approximately true is not due to the fact that the theory fails to approximately fit its domain, but rather due to the fact that the very few approximately true theories are swamped by the very many plainly false. If the theory *is* approximately true, but – due to the correct probabilistic reasoning – we don't believe so, our beliefs will have been led away from the truth. In fact, we may reason as above. Suppose we grant the prevalence of false theories among the successful ones. Then, one might well be predisposed to say that a theory T is false, given its success. When, then, the eyewitnesses (the scientists, in this case) say that a specific theory T is approximately true (despite that this is unlikely, given the base-rates), they should be trusted – at the expense of the base-rates.

The *second* point can be motivated by a certain modification of the Blue Cab/Green Cab example. The situation is as above, with the following difference: the subjects are told that 85 per cent of the car accidents are *caused* by blue cabs and 15 per cent by green cabs. In these circumstances, the subjects did use the base-rates in their reasoning concerning the probability that the culprit was a blue cab (see Koehler 1996, 10). It is easy to see why they did: they thought that the base-rate information, namely, that blue cabs cause accidents much more often than green cabs, was *causally relevant* to the issue at hand. What needs to be emphasised is that in cases such as these there is an *explanation* as to why the base-rate information is relied upon. It's not just because the subjects want to get the probabilities right. It is also because this causally relevant information has a better chance to lead them to true beliefs.

Transpose this case to the problem of truth and success. Suppose there is indeed a high base-rate for false theories. This would be relevant information if it were indicative (or explanatory) of success. If falsity did explain success, then, clearly, the small base-rate for truth would undermine belief in a connection between success and approximate truth. But falsity does *not* explain success. What is more, among the false theories some will be successful and some will be unsuccessful. In fact, it is expected that from a population of false theories, most of them will be *unsuccessful*, while some will be successful. In terms of percentages, it might well be a bit of a fluke that some false theories are successful. The likelihood prob(S/−T) will be low. Actually, it can be so low as to dominate over the high base-rate of false theories. So, suppose that prob(S/−T) = .05, prob(−T) = .9 and prob(S) = .99. Then, prob(−T/S) is .045. A false theory would get no credit at all from success. Conversely, even if the base-rate of truth is low, there is an *explanation* as to why true theories are successful.[11] This might well be enough to show why, despite the low base-rate, a certain successful theory may be deemed approximately true. Its posterior probability may be low, but this will be attributed to the rareness of truth and not to any fault of the individual theory.

Here is another reason why it is, at least occasionally, right to ignore the base-rates. To motivate it, consider again the original Blue Cab/Green Cab case. As above, 85 per cent of the cabs belong to the Green cab company and 15 per cent to the Blue cab one. Imagine that people involved in car accidents are set on taking the cab companies to court. Suppose that on each occasion of the lawsuit, the court takes account of the base-rates and concludes that the cab was green, despite the fact that the eyewitness testified otherwise. Let's say that the court judges that it is always more likely (given the base-rates) that the cab was green (recall that the probability of the cab being blue is .41) and hence it decides to press charges against the Green cab company.[12] If courts acted like that, then the Green company would pay in 100 per cent of such cases, whereas its cabs were responsible for only 59 per cent of such accidents. Fairness and justice seem to give us *some* reason to ignore the base-rates![13]

If we transpose this to the problem of truth and success, the moral should be quite clear. If scientists acted as the imagined judges above, they would be unfair and unjust to their own theories. If, as it happened, the base-rate of false theories were much higher than the base-rate of true ones, they would deem false theories that were true. Conversely, if the base-rate of true theories were much higher than the base-rate of false ones, they would deem true theories that were false.[14]

## 3.6 Taking account of case histories

If we leave base-rates behind, what is left? There are always the case histories to look into. Though it does make sense to raise the grand question 'why is science successful (as an enterprise) as opposed to paradigmatically unsuccessful?', what really matters is the *particular* successes of individual theories, for example, the discovering of the structure of the DNA molecules, or the explanation of the anomalous perihelion of Mercury. If we think of it, it does *not* matter for the truth of the double-helix model that truth is hard to get. The base-rate of truth (or of falsity) – even if we can make sense of it – is outweighed by the case history. We have lots of detailed information about the DNA-molecule case to convince us that the double-helix model is approximately true, even if, were we to factor in the base-rate of true theories, the probability of this model being approximately true would be very low. We are right in this case to ignore the base-rate, precisely because we know that this model's being approximately true does *not* depend on how many other true or false theories are around.

This last observation seems to me quite critical. The approximate truth of each and every theory will *not* be affected by the number (or the presence) of other theories (even more so if those are independent of the given theory). Approximate truth, after all, is a relation between the theory and its domain (a relation of approximate fit). This relation is independent of what other (true or false) theories are available. In fact, we can see there is an ambiguity in the probabilistic formulations of NMA. Though I have hinted at this above, it is now time to make it explicit.

There are *two* ways to think of arguments such as (A) and (B) above. The first is to apply the argument to a *specific* theory T (say, the electron theory, or Newtonian mechanics or the special theory of relativity). Then we ask the question: how likely is this *specific* theory T to be true, given that it has been successful? The second way is to apply the argument to an *arbitrary* theory T. Then we ask the question: how likely is an *arbitrary* (randomly selected) theory T to be true, given that it has been successful? If the issue is posed according to this second way, it does follow from Bayes's theorem that the probability of a theory's being approximately true will depend on (and vary with) the base-rate of true theories. But if the issue is raised for a specific theory, then base-rates have no bite at all. Even if we had the base-rates, there are good reasons to neglect them – and scientists do neglect them – when the case history offers abundant information about the approximate truth of a given theory.[15] (A related but different approach to this issue is advanced in Chapter 4, Section 4.4, where a distinction is drawn between two kinds of evidence in favour or against scientific theories.)

## 3.7 Likelihoodism

We are not done yet. The subjective Bayesian might now come back with a vengeance. He might say: ditch the base-rates, and go for purely subjective estimates of how likely it is that a theory is true. Consider what Howson (2000, 58) says: '(F)ar from showing that we can ignore even possibly highly subjective estimates of prior probabilities, the consideration of these quantities is *indispensable* if we are to avoid fallacious reasoning.' So, can we do away with priors altogether? Recall the Bayes factor from Section 3.4. As Kevin Korb (2002, 4) has argued, this factor reports the 'normative *impact* of the evidence on the posterior probability, rather than the posterior probability itself'. To get the posterior probability, we also need the prior. If the Bayes factor $f$=prob(S/–T)/prob(S/T)=1, then prob(T/S)=prob(T), that is, the success of a theory makes no difference to its truth or falsity. But, the further from unity $f$ is, the greater is the *impact* of the evidence. If $f$=0, prob(T/S)=1. And if $f$ tends to infinity, then, given that prob(T)>0, prob(T/S) tends to 0. Given all this, it seems that we can reformulate Howson's NMA ($B^1$) in such a way that it avoids base-rates (prior probabilities). The idea is that NMA need not tell us how probable a theory is, given the evidence (or its success). Rather, it tells us what the impact of the evidence (or the success) is on the posterior probability of the theory (without assuming that there is need to *specify* this posterior probability, and hence need to rely on a prior probability).

($B^4$)
*f* is close to zero (i.e., prob(S/–T) is close to zero and prob(S/T) is close to 1).
S is the case.
Therefore, the impact of S on prob(T/S) is greater than its impact on prob(–T/S).

($B^4$) can be supplemented with some specification of prior probabilities and hence it can yield a concrete posterior probability. Thus, it can become either ($B^2$) or ($B^3$) above. But, even as it stands, it is suitable for *modest* Bayesians, who just want to capture the comparative impact of the evidence on competing hypotheses.

We should also take a look at what has been called 'likelihoodism'. As Elliott Sober (2002) understands it, likelihoodism is a modest philosophical view. It does not aim to capture all epistemic concepts. It uses the *likelihood ratio* to capture the strength by which the evidence supports a hypothesis over another, but does not issue in judgements as to what the probability of a hypothesis in light of the evidence is. In particular, likelihoodism does not require the determination of prior probabilities. It does not tell us what to believe or which hypothesis is probably true. Given two hypotheses $H_1$ and $H_2$, and evidence *e*, likelihoodism tells us that *e* supports $H_1$ more than $H_2$ if $prob(e/H_1) > prob(e/H_2)$. The likelihood ratio $prob(e/H_1)/prob(e/H_2)$ is said to capture the strength of the evidence.

Note that the likelihood ratio $f^* = prob(e/H_1)/prob(e/H_2)$ is the converse of the Bayes factor *f*, as defined above. So likelihoodists can adopt a variant of ($B^4$):

($B^5$)
$f^*$ is greater than one (i.e., prob(S/T) is close to 1 and prob(S/–T) is close to zero).
S is the case.
Therefore, S supports T over –T.

It is not my aim presently to defend either ($B^4$) or ($B^5$). But it should be stressed that if we have in mind a more modest version of NMA, that is, that success tells more strongly in favour of truth than of falsity, then we can take ($B^4$) as a version of NMA suitable for modest Bayesians and ($B^5$) as a version of NMA suitable for non-Bayesians.[16]

Let me recapitulate: The moral of Sections 3.2 and 3.3 is that there is no reason to think of the Ultimate Argument for realism as a deductive argument, contrary to what Musgrave suggests. The moral of Sections 3.4–3.7 is that we should also resist the temptation to cast the NMA in a(n) (immodest) subjective Bayesian form. Once we free ourselves from both deductivism and subjective Bayesianism, there is no reason to think that NMA is either deductively or inductively fallacious. Many will remain unpersuaded. Both deductivism and Bayesianism are all-encompassing (shall I say imperialistic?) approaches to reasoning and have many attractions (and a number of well-known successes). In fact, they share a common central theme: reasoning has a certain formal structure (given by deductive rules and Bayes's theorem – or better Bayesian conditionalisation). The substantive assumptions that are employed in reasoning have to do either with the truth of the premises (in deductivism) or with the prior probabilities (in Bayesianism). But perhaps, the simplicity of both schemes of reasoning is their major weakness. Reasoning is much more complex than either of them admits. (This point is elaborated in Psillos 2007a.)

So, what sort of argument is the Ultimate Argument for realism? I know of no more informative answer than this: it is an IBE. And what kind of inference is IBE? I know of no more informative answer than this: it is the kind of inference which authorises the acceptance of a hypothesis H as true, on the basis that it is the best explanation of the evidence. The rationale for IBE is that explanatory considerations should inform (perhaps, determine) what is reasonable to believe. All this is still too crude to count as an explication, but some of the details are offered in Chapter 10.

# 4
# Against Neo-Instrumentalism

Stanford (2006) has advanced a sophisticated neo-instrumentalist alternative to scientific realism. In fact, he has gone as far as to suggest that it may be a mistake to try to identify 'a crucial difference' between instrumentalism and realism when it comes to the epistemic attitudes they recommend towards theories or theoretical assertions. There is, he says, at most 'a local difference in the specific theories each is willing to believe on the strength of the total evidence available' (ibid., 205). Though I welcome this attempt at reconciliation, I will argue that Stanford's own way to achieve it, while keeping a distinguished instrumentalist outlook, is flawed.

## 4.1 New induction vs pessimistic induction

Stanford starts with a bold claim, namely, that at any given stage of inquiry there have been hitherto unconceived but radically distinct alternatives to extant scientific theories. When, in the fullness of time, these alternatives came to be formulated, they were equally well confirmed by the then available evidence; they came to be accepted by scientists in due course; and eventually they replaced their already existing rivals. This is a condition that he calls 'Recurrent Transient Underdetermination'. If theories are subject to this predicament, Stanford argues, belief in their truth is not warranted. Not all theories are indeed subject to this, but Stanford thinks that all fundamental scientific theories in a variety of domains of inquiry suffer from recurrent radical underdetermination by the evidence. Based on evidence coming from the history of science, he performs what he calls the *New Induction* (NI): there are good inductive reasons to believe that for *any* fundamental theory scientists will come up with – and for *any* evidence that will be available – there will be hitherto unconceived theories that will be at least as well confirmed as the ones available. This kind of situation is supposed to be the springboard for breathing new life into instrumentalism. It promises to show that (there are good reasons to believe that) fundamental theories are not accurate descriptions of the deep structure of the world but rather 'powerful conceptual tools for action and guides to further inquiry' (Stanford 2006, 24–5).

Suppose, for the sake of the argument, that we grant all this. It should be immediately noted that realism about fundamental theories would be in jeopardy only if the Pessimistic Induction (PI) were sound. The NI can only work in tandem with the PI. Unless PI is correct, NI does not suffice to show that the new and hitherto unconceived theories will be radically dissimilar to the superseded ones. Hence, rehabilitating PI is an important step in Stanford's strategy.

## 4.2 Resisting PI's rehabilitation

Recall from Chapter 1, Section 1.2 that recent realist responses to PI have aimed to show that there are ways to distinguish between the 'good' and the 'bad' *parts* of past abandoned theories and that the 'good' parts – those that enjoyed evidential support, were not idle components and the like – were retained in subsequent theories. This kind of response aims to show that there has been enough theoretical continuity in theory-change to warrant the realist claim that science is 'on the right track'. This kind of response damages (at least partly) Stanford's unconceived alternatives gambit. If there is convergence in our scientific image of the world, the hitherto unconceived theories that will replace the current ones won't be the radical rivals they are portrayed to be. Claiming convergence does not establish that current theories are true, or likely to be true. Convergence there may be and yet the start might have been false. But the convergence in our scientific image of the world puts before us a candidate for explanation. The generation of an evolving-but-convergent network of theoretical assertions is best explained by the assumption that this network consists of approximately true assertions.

Stanford's main objection to this way to block PI is that it is tailor-made to suit realism. He claims that it is the fact that the very same present theory is used *both* to identify which parts of past theories were empirically successful *and* which parts were (approximately) true that accounts for the realists' wrong impression that these parts coincide. Stanford (2006, 166) says:

> With this strategy of analysis, an impressive retrospective convergence between our judgements of the sources of a past theory's success and the things it 'got right' about the world is virtually guaranteed: it is the very fact that some features of a past theory survive in our present account of nature that leads the realist *both* to regard them as true *and* to believe that they were the sources of the rejected theory's success or effectiveness. So the apparent convergence of truth and the sources of success in past theories is easily explained by the simple fact that both kinds of retrospective judgements have a common source in our present beliefs about nature.

I find this kind of objection misguided. The way I see it, the problem is like this. There are the theories scientists currently believe (or endorse – it does not matter) and there are the theories that were believed (endorsed) in the past. Some (but not all) of them were empirically successful (perhaps for long periods of time). They were empirically successful irrespective of the fact that, subsequently, they came to be replaced by others. This replacement was a contingent matter that had to do with the fact that the world did not fully co-operate with the extant theories: some of their predictions failed; or the theories became overly ad hoc or complicated in their attempt to accommodate anomalies, or what have you. The replacement of theories by others does not cancel out the fact that the replaced theories were empirically successful. Even if scientists had somehow failed to come up with new theories, the old theories would not have ceased to be successful. So success is one thing, replacement is another. Hence, it is one thing to inquire into what features of some past theories accounted for their success and quite another to ask whether *these* features were such that they were retained in subsequent theories of the same domain. These are two independent issues and they can be dealt with (both conceptually and historically) independently of each other.

They can be mixed up, of course. A (somewhat) careless realist could start with current theories and then try to tell a story about the abandoned and replaced ones such that it *ensures* that some of the theoretical assumptions about the world that scientists currently endorse were present in the past theories *and* responsible for their empirical successes. But carelessness is not mandatory! One can start with some past theories and try on independent grounds – bracketing the question of their replacement – to identify the sources of their empirical success; that is, to identify those theoretical constituents of the theories that fuelled their successes. This task won't be easy, but there is no principled reason to think it cannot be done. Unless, of course, one thinks that when a prediction is made the whole of the theory is indiscriminately implicated in it – but this kind of blind holism is no more than a slogan, or a metaphor. When a past theory has been, as it were, anatomised, we can *then* ask the independent question of whether there is any sense in which the sources of success of a past theory that the anatomy has identified are present in our current theories. It's not, then, the case that the current theory is the common source for the identification of the successful parts of a past theory *and* of its (approximately) true parts. Current theories constitute the vantage point from which we examine old ones – could there be any other? – but the identification of the sources of success of past theories need not be performed from this vantage point.

What needs to be stressed is that the realist strategy proceeds in two steps. The first is to make the claim of convergence plausible, namely, to show that there is continuity in theory-change and that this is not merely empirical continuity; substantive theoretical claims that featured in past theories and played a key role in their successes (especially novel predictions) have been incorporated (perhaps somewhat re-interpreted) in subsequent theories and continue to play an important role in making them empirically successful. This first step, I take it, is common place – unless we face a conspiracy of the scientific community to make us believe that every time a new theory is advanced and endorsed scientists do not start from square one (though they actually do). As noted above, this first step does not establish that the convergence is to the truth. For this claim to be made plausible a second argument is needed, namely, that the emergence of this stable network of theoretical assertions is best explained by the assumption that is, by and large, approximately true. The distinctness of these two steps shows that Stanford's criticism is misguided.[1]

## 4.3 Exhausting the conceptual space: No thanks

What is the evidence for the NI? What is Stanford's reference class? How exactly are theories individuated? Consider, as an example, gravitational theories. One could argue that, if they are understood very broadly, there have been just three of them: Aristotelian, Newtonian and Einsteinian. Is the inductive evidence of two instances enough to warrant Stanford's inductive conclusion? Alternatively, one could argue that though there is a genuine shift from Aristotelian to Newtonian theories, the shift from

Newtonian to Einsteinian is, by hindsight, not too radical concerning the space of possibilities. Since Newtonian theories can be reformulated within the general four-dimensional framework of General Relativity, there is a sense in which the conceptual space of possibilities of Newton's overlaps with the conceptual space of General Relativity. In any case, one could argue that as science grows, theories acquire some stable characteristics (they become more precise; the evidence for them is richer and varied; they are more severely tested; they are incorporated into larger theoretical schemes and others) such that (a) they can no longer be grouped together with older theories that were much cruder or underdeveloped to form a uniform inductive basis for pessimism and (b) they constrain the space of alternative possibilities well enough to question the extent of the unconceived alternatives predicament.

In a rather puzzling move, Stanford tries to neutralise these worries by shifting his attention from theories to *theorists*. Stanford (2006, 44) claims that his arguments intend to show that 'present theorists are no better able to exhaust the space of serious well-confirmed possible theoretical explanations of the phenomena than past theorists have turned out to be.' There is a sense in which this kind of claim is question-begging. It assumes that the conceptual space of theories that can possess several empirical and theoretical virtues is *inexhaustible*, or that there has been no progress at all in delimiting it substantially. I see no argument for these considerations except for the idea that every time scientists come up with a new theory, they start from scratch. But this would simply beg the question. The NI in itself does not imply that the conceptual space of theories is inexhaustible; rather, if its conclusion is to be warranted at all, it must be presupposed that the conceptual space of theories is inexhaustible. At best, at this point we reach a stand-off. Stanford needs an independent argument to show that the conceptual space of theories is inexhaustible and the realist needs an independent argument that every new theory in science is not a false start. There is no advantage on either side. But let's leave all this to the one side.

Focusing the issue on theorists – on human scientists, that is on cognitive creatures of the sort we are – is particularly unfortunate. Suppose Stanford's claim is true. Suppose that scientists have been poor conceivers of alternatives of extant scientific theories. What would this show? It would report an interesting psychological fact: something about our cognitive faculties, or constitution. Dogs are not expected to understand high-level theories; this is a fact of life for them (I presume). But we do not reach this important conclusion, we do not arrive at this important fact about dogs, by performing an induction on their past *failure* to understand high-level theories. Even if we use some evidence for this cognitive shortcoming that comes from the past performances of dogs, we aim for a theoretical explanation of this fact – we study their cognitive life not because we want to improve it (thought we might) but because we want to understand why they have the limitations they do. We are not like dogs (in this respect) – we think we understand a lot about the workings of nature and that we have made progress in this understanding. We even have theories as to how this understanding is possible. Would (or should) we change all this on the basis of the NI? Would we base a severe cognitive limitation on such flimsy evidence? Shouldn't we require an explanation and/or account of this fundamental cognitive shortcoming?

Any decent explanation of how this cognitive limitation occurs would rely heavily on theories of the sort Stanford thinks there are unconceived alternatives. Hence, by Stanford's lights, we would not even be entitled to such an explanation – we would have to base a great fact about us and our cognitive prospects on a simple enumerative induction which faces obvious problems, especially when there is no clear reference class as its basis. I do not want to buttress enumerative induction. But let us think of the following situation. Stanford invites us to accept a fundamental cognitive shortcoming on the basis of an enumerative induction. In effect, he invites us to distrust the eliminative methods followed by scientists because they (*qua* theorists) 'are not well suited to exhausting the kinds of spaces of serious alternative theoretical possibilities' (Stanford 2006, 45). But hasn't Stanford undermined himself? Is he not in the very same predicament? It might be the case that he occupies a position in which there is some serious (and hitherto unconceived) alternative explanation of the fact that scientists have tended to occupy a position in which there are unconceived, but at least as well-confirmed, theories. In this case, Stanford is not entitled to endorse his own conclusion. The right attitude, here, is not to shift the burden of proof to the opponents, as Stanford (cf. 2006, 46) does. To make his claim credible, Stanford must show that he is not subject to the kind of predicament that other theorists are supposed to be. It is not enough to say that we do not have 'any clear reason to distrust' the conclusion of the NI – we need a positive reason to trust it and this, by Stanford's own lights, must be such that the space of alternative hitherto unconceived explanations has been exhausted.

Let us pretend that it is an established fact the scientists are poor conceivers of alternative theories. If we were to be told of this fact outside the confines of the realism debate, the first thing all of us would insist on is that we must *do* something about it. When people are poor in solving mathematical problems, they should be better trained. In a similar fashion, we (i.e., cognitive scientists or whoever else is interested) should try to improve the scientists' ability to conceive alternatives – perhaps by having classes of creative imagination where they devise radical alternative theories at random and then test them severely.

Leaving this half-joke aside, the problem of unconceived alternatives – if it is a problem at all – is not a problem for realism in particular. It is a general problem for epistemology. Given that there is no deductive link between evidence and theory, it is always possible that there are unconceived alternatives of extant theories that go beyond the evidence – no matter by how little. The issue, then, is under what conditions we are entitled to talk about justification and knowledge when proof is not available. This is a normative issue and should be dealt with independently of the realism debate. No-one would seriously maintain (nowadays at least) that justification (and knowledge) requires deductive proof; or elimination of all but one possible theories or explanations. Even Stanford talks about *serious* and *plausible* alternatives. He takes it that the problem of unconceived alternatives does not pose problems in inferences in which the 'grasp of the plausible alternatives is indeed exhaustive' (Stanford 2006, 38). Though more may be said on all this, suffice it for the moment to say that no-one has a privileged access to what the plausible alternatives are and when (and whether) they have been exhausted. It's best if we leave this kind of judgement to scientists. If we criticise them on the grounds that they are poor conceivers of plausible alternatives, we had better have a good theory of what makes an alternative plausible and when the plausible alternatives have (or have not) been exhausted.

## 4.4 Balancing two kinds of evidence

The fact is that when we think about scientific theories and what they assume about the world we need to balance two kinds of evidence. The first is whatever evidence there is in favour (or against) of a specific scientific theory. This evidence has to do with the degree on confirmation of the theory at hand. It is, let us say, *first-order evidence*, say about electrons and their having negative charge or about the double helix structure of the DNA and the like. First-order evidence is typically what scientists take into account when they form an attitude towards a theory. It can be broadly understood to include some of the theoretical virtues of the theory at hand (parsimony and the like) – of the kind that typically go into plausibility judgements associated with assignment of prior probability to theories. These details do not matter here. (But see the relevant discussion in Chapter 5, Section 5.4 and Chapter 10, Sections 10.3.1 and 10.3.2.) The relevant point is that first-order evidence relates to specific scientific theories and is a major determinant of judgements concerning the confirmation of the theory. The second kind of evidence (let's call it *second-order evidence*) comes from the past record of scientific theories and/or from meta-theoretical (philosophical) considerations that have to do with the reliability of scientific methodology. It concerns not particular scientific theories but science as a whole. (Some) past theories, for instance, were supported by (first-order) evidence, but were subsequently abandoned; or some scientific methods work reliably in certain domains but fail when they are extended to others. This second-order evidence feeds claims such as those that motivate the PI, or the NI, or (at least partly) the argument from the underdetermination of theories by evidence. Actually, this second-order evidence is multi-faceted – it is negative (showing limitations and shortcomings) as well as positive (showing how learning from experience can be improved). The problem, a *key* problem I think, is how these two kinds of evidence are balanced.

And they need to be balanced. For instance, the way I see it, the situation with the PI is this. There is first-order evidence for a belief *P* (or a theory T). At the same time, there seems to be second-order evidence for the (likely) falsity of a belief corpus *S*, which includes *P*. The presence of second-order evidence does not show that *P* is false. Yet, it is said to undermine the warrant for *P* (by undermining the warrant for the corpus in which *P* is embedded). The question then is: are there reasons for *P* that override the second-order evidence against *S*, which in its turn undermines *P*? This cannot be settled in the abstract – that is without looking into the details of the case at hand. To prioritise second-order evidence is as dogmatic as to turn a blind eye to it. My own strategy in (Psillos 1999) has been to show that there can be first-order reasons for believing in *P* (based mostly on the *P*'s contribution to the successes of theories) that can defeat the second-order reasons the PI offers against *P* – where the *P*'s are certain theoretical constituents of past theories.

Critics of realism have tended to put a premium on the negative second-order evidence. But one has to be careful here. An uncritical prioritisation of second-order evidence might well undermine the intended position of the critic of realism. The second-order evidence feeds the very same methods that scientists have employed, based on first-order evidence, for the assessment of their theories. By drawing a conclusion that undermines the credibility of scientific theories, it may well undermine the credentials of this very conclusion. Take for instance the PI: it might be seen as a challenge to the reliability of the method employed to warrant belief in certain propositions, say *P*. But then the very conclusion of PI is undermined if it is taken to be based on the very method whose reliability PI has undermined.

In Stanford's case, the premium is put on the inductive evidence that scientists are poor conceivers of alternative explanations. Actually, the premium seems so high that Stanford disregards the first-order evidence there is in favour of scientific theories. More accurately put, Stanford allows this evidence to be taken into account *only* in the cases where (it is justifiedly believed that) there are no plausible alternatives. This may well have things the wrong way around. The justification of the belief that it is unlikely that there are (plausible) alternatives to a given theory is surely a function of the degree of confirmation of theory. It is only by totally disregarding this first-order evidence that Stanford is able to make the justification of the belief about how likely it is that there are (plausible) alternatives to a given theory solely a function of whatever second-order evidence there is for or against the conceivability record of scientists. One can easily envisage the following situation. Evidence *e* supports theory *T*. Hitherto unconceived theory $T'$ is at least as well confirmed as *T*. $T'$ is formulated and replaces *T*. $T'$ is further confirmed by evidence $e'$. The support of $T'$ is so strong ($T'$ is so well confirmed) that it becomes pointless to question it and to envisage serious and plausible unconceived rivals. To claim that this kind of scenario is *unlikely* is to claim either that theories are not really confirmable by the evidence or that their degree of confirmation can never be strong enough to outrun whatever evidence there is for the poor conceivability predicament. Both claims are implausible. In fact, the (first-order) evidence for a theory might be so diverse and varied that, by anyone's methodological standards, it supports the theory more strongly than the (second-order) evidence that supports the poor conceivability predicament.

## 4.5 Liberal instrumentalism

Stanford's instrumentalism, we have already noted, is sophisticated and liberal. Stanford accepts that predictions are theory driven and that they involve theoretical descriptions of whatever is predicted – being observable or unobservable. He puts no special epistemic weight on the observable–unobservable distinction. He takes it that our understanding of the world is theoretical 'all the way down' (Stanford 2006, 202);

that theories are our best conceptual tools for thinking about nature (cf. 2006, 207). In fact, his point of view is not instrumentalism *tout court*. According to his core characterisation of neo-instrumentalism, theories are predictive and inferential tools, but the inferences they licence – relying indispensably on theoretical descriptions – are not from observables to observables but 'from states of affairs characterised in terms we can strictly and literally believe to other such states of affairs' (Stanford 2006, 203–4). In other words, Stanford's instrumentalism relies on a body of *strict and literal beliefs* that form the basis on which the reliability of the theories as instruments for inference and prediction is examined. This body should, emphatically, *not* be associated with more traditional instrumentalist commitments to observables or to sensations and the like. How is it, then, to be circumscribed?

Here is where some back-pedalling starts. Instrumentalists have always taken it to be the case that no matter how attractive and useful theories might be, their 'cash value' has to do with what they say about the macro-world of experience. This is, in essence, what Edmund Husserl called the 'life-world', which he took it to be the 'pregiven' world we live in, that is, the 'actually intuited, actually experienced and experienceable world' (Husserl 1970, 50). The content of this world might be understood narrowly or broadly (as Husserl understood it). Be that as it may, the point is that the content of the life-world is supposed to be accessed independently of theories. This is very similar to what Stanford thinks. He takes it that there is a part of the world to which there is 'some independent route of epistemic access' (Stanford 2006, 199). To be more precise, Stanford claims that some parts of the world can be understood in terms of theory towards which there can be no instrumentalist stance; these parts of the world (so characterised) will be the benchmark against which the reliability of the instrumentally understood theories is checked. Note that this talk of independent epistemic access does not imply theory-free access. It implies theory-driven access, but by means of theories that are (believed to be) true. This, Stanford thinks, allows us to identify those 'concrete consequences of a theory [that] can be grasped independently of the theory' and hence to 'specify the beliefs about the world to which we remain committed once we recognize that we are not epistemically entitled to believe in a particular theory' (ibid., 198).

Stanford is aware that for this line of argument to work, the theories that provide the independent route of epistemic access to some parts of the world should *not* be subject to the unconceived alternatives problem. So he proceeds in two steps. The first is to argue that 'the hypothesis of bodies of common sense', namely, the common sense ontology of middle-sized physical objects, is immune to the unconceived alternatives predicament. This is supposed to be so because there is no historical record of finding serious unconceived alternatives to it. The second step is to argue that this 'hypothesis of bodies of common sense' is not, in all probability, the only serious sound theory that will remain available to the instrumentalist. More specifically, Stanford tries to show that several low- or middle-level theories that involve a good deal of explanatory sophistication won't be challenged by unconceived alternatives.

This kind of liberalisation of instrumentalism renders it, I think, more credible. In a way, it makes the realist-instrumentalist debate more local and less polarised. The focus is shifted to *particular* theories and to whether or not they fall victims to the unconceived alternatives problem.

Having said this, let me state a number of problems that Stanford's liberal instrumentalism faces. *First*, it is unfortunate that it is based on the presence of a body of strict and literally true beliefs. I doubt there are any such beliefs. Even if there are, they are not very interesting. Most beliefs of the common sense – those that are supposed to form the backbone of the independent route of epistemic access to the part of the world the instrumentalist is interested in – are neither literally true, nor strict and precise. Is the surface of the pool table flat? Well, it depends. Is the height of John 1.73? Close enough. Is the earth round? FAPP. Is the sea water blue? Not quite. Is whale a fish? Not really. Do unsupported bodies fall to the ground? Yes, but... Does aspirin cause headache relief? It's very likely. This is just a figurative way to make the point. And the point is that common sense is *not* a theory towards which we can have a stance of strict and literal belief. Many terms and predicates we use in our commonsensical description of the world are vague and imprecise. As is explained in some detail in the next chapter, Section 5.3, gaining independent epistemic access to the very entities assumed by the common sense requires leaving behind (at least partly) the common-sense framework. *Second*, is the hypothesis of common bodies immune to the problem of unconceived alternatives? Not really. Let's forget about phenomenalism and the like – for the sake of the argument. The biggest – at some point unconceived – alternative to the hypothesis of common bodies is science itself. Well, we might not think of science as a rival to common sense. But it is not exactly continuous with it either. Science revises some of the categories by means of which we think of common bodies. The table is not solid (at least, there is a well-defined sense in which it is not); colours are not irreducible phenomenal qualities and the like. Arguably, the scientific framework has replaced the framework of the common sense. *Third*, Stanford gets the contrast wrong. The contrast is not common bodies vs calorific molecules, circular inertia, electrons, strings and the like. The contrast should be between common bodies and theoretical entities (meaning: assumed constituents of common bodies). And the relevant point is that the hypothesis of theoretical entities has no serious and plausible unconceived alternatives. Hence, by Stanford's own light, this should be quite secure.

There is a more general problem with Stanford's strategy to which I will focus my attention. Suppose he is right in what he says about the hypothesis of common bodies and the network of strict and literally true beliefs we have concerning common bodies. Stanford's strategy is overly conservative. It favours rather elementary theories. The irony is that it is *known* that the favoured theories are elementary, for otherwise there would be no motivation (acknowledged by Stanford's instrumentalism too) to advance more sophisticated theories (like proper scientific theories) so as to improve our understanding of the very objects assumed by the elementary theories. There is an issue of *motivation*, then: why try to devise theories? If the common sense framework is already in place and consists of a body of (strictly and literally true beliefs), why not stay there? The answer, of course, is that we know that this framework admits of corrections and that scientific theories correct it (as well as their own predecessors). This is now commonplace: science is not just piled upon common sense. It adds to it and it corrects it. What perhaps Husserl did not consider when he aimed to show the priority of the life-world over the scientific image, Stanford and everybody else by now have taken to heart (cf. 2006, 201). But then there is an issue of *explanation* – which I will phrase in very general terms: if newer theories correct the observations, predictions, commitments and the like of their predecessors, they cannot just be more reliable instruments than their predecessors – this would not explain *why* the corrections should be trusted and be used as a basis for further theoretical developments.

This last line of thought might be pressed in several ways, but the most promising one in this context seems to be the following. Assume there is a network of strictly and literally true beliefs on which we start our endeavours. As science grows, there are two options. The first is that this network does not change – no more strictly and literally true beliefs are added to it as a result of the scientific theorising, theory testing and the like. This would render bells and whistles all this talk about the indispensability of scientific theorising. The whole spirit of Stanford's liberal instrumentalism would be violated. The second option (rightly favoured by Stanford himself) is that the network of strictly and literally true beliefs is enlarged – more stuff is added when new theories are advanced, accepted, tested and the like. Recall, that Stanford holds no brief for the observable–unobservable distinction and does not restrict his realist stance to simple empirical laws and generalisations. What happens then? Every new theory will enlarge the domain that is interpreted realistically (i.e. it is the subject of literal and strict belief). So every successor theory will have more realistic (less instrumentally understood) content than its predecessor. As this process continues (or has continued in the actual history of science), one would expect that at least some parts of the theory that Stanford treats as instrumentally reliable will become so involved in the interpretation of the realistic parts of the theory that it won't be cogent to doubt them without also doubting the realistically interpreted parts of the theory. I take it that this is the case with the atomic hypothesis, nowadays. But the point is very general. And it is that this kind of neo-instrumentalist image of science might well undermine itself and allow realism to spread indefinitely.

Stanford might well claim that realism will never spread to high-level and fundamental theories. He does admit that there are cases in which the available evidence constrains the space of competing available explanations – he argues, for instance, that the possibility that amoebas do not exist is ruled out. But he also insists that there are contexts – mostly having to do with fundamental physics – in which the evidence will never rule out alternative competing explanations. Perhaps, what has already been said in relation to the unconceived alternatives predicament is enough to make it plausible that there is no principled difference between being committed, say, to the reality of amoebas and being committed to the reality of atoms. If scientists are poor conceivers, why should we think they can succeed with amoebas but not with atoms?

The case of atoms is particularly instructive. Peter Achinstein (2002) has argued that irrespective of the prospects of finding a general argument for realism (like the no-miracles argument discussed in Chapter 3), there is plenty of evidence to accept the reality of *atoms*. This has come mostly from the work of the experimentalist Jean Perrin. Part of Achinstein's strategy has been to show that Perrin did consider and eliminate a number of alternative potential explanations of certain phenomena. This strategy capitalises on what above I called 'first-order evidence' for a scientific theory. Clearly, part of the weight of the evidence in favour of a theory is carried by the process of eliminating alternative potential explanations. Stanford's point against Achinstein is that we can *reasonably* doubt Perrin's hypothesis because we might doubt his ability to exhaust the space of alternative explanations, 'without being able to specify a particular alternative that he has failed to consider' (Stanford 2006, 40). As noted already, this kind of claim is so sweeping (it talks about *exhaustion* of the conceptual space and it considers it enough that there *might* be alternative explanations, even if we cannot specify any), that it undercuts any claim to knowledge whatever – common bodies and amoebas and dinosaurs go the way of Perrin's atoms.

Ultimately, Stanford's instrumentalist – like many other instrumentalists – relies on double standards in confirmation. The typical strategy here is this: the content of a theory is split into two parts – let's call them the OK-assertions and the not-OK-assertions, respectively. This partition can be made along several dimensions, but typically it is made along the lines of empirical vs theoretical or observable vs unobservable. The OK-assertions are said to be confirm*able* and confirmed by the evidence. Then the further claim is made (or implied) that the principles of confirmation that concern the OK-assertions are *not* transferable to the not-OK-assertions. Stanford seems to be in a worse situation here because he allows (in fact he requires) that some theories (and some claims about unobservables) are strictly and literally believed. But then he has to show that the ways in which the beliefs he allows (the OK-assertions) are confirmed are radically different from the ways in which the non-OK assertions confront the relevant evidence. No such case has been made. In fact, the very motivation for double standards in the theory of confirmation is deeply problematic. There is plenty of reason to think that the very same principles and methods are implicated in the confirmation of both the OK-parts and the not-OK parts. In other words, the very distinction between the OK-parts and the not-OK parts of a scientific theory is suspect.

Is there room for rapprochement between liberal versions of instrumentalism and scientific realism? I think there is, subject to two constraints. First, there should not be double standards in confirmation, based on a supposedly principled distinction between OK-entities and not-OK ones. Second, there should be no absolute privilege conferred on either the first-order or the second-order evidence that is brought to bear on scientific theories. Given these two constraints, a variety of stances towards scientific theories might be enunciated, depending on the context – broadly understood.

There may be independent and different reasons for treating a theory unrealistically. A theory might be known to be radically false in that there is massive evidence against its basic posits and explanatory mechanisms – for example, the Ptolemaic theory. A theory might be known to be false, but it might be such that it gets some really fundamental explanatory mechanisms right (e.g., the Copernican theory). A theory might be known to be false, even in some fundamental explanatory mechanisms, but it might have many constituents that play important predictive and explanatory role and have not be shown to false – at least they have been accommodated within subsequent theories (e.g., the Newtonian theory). It might be that a theory is known to be false, not because it gets some fundamental explanatory mechanisms wrong, but because (as is typically the case) it involves idealisations and abstractions – in this case, the theory might be approximately true or by-and-large right if there are determinable and quantifiable respects and degrees in which real-worldly entities match up the theory's idealised posits. There is no reason to have a blanket attitude towards all these types of theories.

The degree of credence attributed to a scientific theory (or to some parts of it) is a function of its degree of confirmation. This requires balancing the first-order evidence there is in favour of a theory with second-order considerations that have to do with the reliability of scientific method, the observed patterns in the history of science and the like. The weight of the second-order evidence is not necessarily negative. Science is not a random walk. The static picture of scientists' starting afresh each time a theory is revised or abandoned should be contrasted to a *dynamic* image, according to which theories improve on their predecessors, explain their successes, incorporate their well-supported constituents and lead to a more well-confirmed (and according to current evidence, truer) description of the deep structure of the world.

If we abandon philosophically driven dichotomies imposed on the body of scientific theories, we can take science to be a dynamic process which, on balance, has been quite successful in disclosing the deep structure of the world.

# 5
# Tracking the Real: Through Thick and Thin

Azzouni (2004) invites us to consider the way we ought to form beliefs about what we take to be real. Beliefs are the products of epistemic processes (e.g., observation or inference) but the processes Azzouni recommends should meet his 'tracking requirement': they should be 'sensitive to the objects *about which* we claim to be establishing (the truths we are committed to)' (Azzouni 2004, 371–2). Azzouni (2004; 1997) claims that the tracking requirement is met by what he calls 'thick epistemic access' and, in particular, *observation*. Thick epistemic access is defined as follows:

> Any form of epistemic access which is robust, can be refined, enables us to track the object (...), and which (certain) properties of the object itself play a role in how we come to know (possibly other) properties of the object is a *thick* form of epistemic access. (Azzouni 1997, 477)

To this kind of access to the real (*thick*), he contrasted what he called 'thin' epistemic access. This is taken to be access to objects and their properties via a *theory*, which is (holistically) confirmed and has the Quinean virtues (simplicity, familiarity, scope, fecundity and success under testing). The friends of thin access would have it that 'if a theory has these virtues, we have good (epistemic) reasons for adopting it, and all the posits that come with it' (ibid., 479).[1] But Azzouni is no friend of thin epistemic access. His prime claim is that 'thin epistemic access' fails to secure commitment to the real.

In this chapter, I examine critically Azzouni's tracking requirement and its use as a normative constraint on theories about objects that we take as real.

## 5.1 Theoretical irrealism vs holistic realism

Azzouni aims to occupy a middle position between what he calls 'theoretical irrealism' and, what one may call, *holistic realism*. Holistic realism relies on confirmational holism: insofar as a theory is confirmed by the evidence as a whole, and is taken literally, ontic commitment accrues to whatever is posited by this theory – and hence to its unobservable posits. Quine's views are taken to be the *locus classicus* of holistic realism. According to Azzouni, far from being realist, this view slides towards 'idealism' (Azzouni 2004, 377). There is no reason, he thinks, to believe that thin epistemic access, via theory-confirmation, meets the tracking requirement. The Quinean virtues of well-confirmed theories fail 'to track the properties of the objects such theories are about' (ibid., 378). 'Theoretical irrealism', on the other hand, *denies* the existence of unobservable entities. It bases this denial on the claim that while some epistemic processes (those that rely essentially on observations) track observables and their (observable) properties, other epistemic processes (those that rely essentially on the confirmation of theories) *fail* to track unobservables (if there are any) and their properties. Azzouni's theoretical irrealist honours the tracking requirement but restricts it to observable entities and the epistemic processes which rely on observation.

Here is where Azzouni himself comes in. Insofar as he thinks that confirmational holism does not meet the tracking requirement, he is in full alignment with the theoretical irrealist. Yet, where the theoretical irrealist *stops* at the claim that observable entities are real because they are capable of being thickly tracked, Azzouni argues that the tracking requirement can be met by (instrument-based) epistemic processes that give access to at least *some* unobservables. So, Azzouni intends to block theoretical irrealism and to defend some form of scientific realism. How is this? As we have seen, Azzouni takes observation to be an exemplar of thick access to the real. Azzouni (1997, 477) says: 'all observations of something are thick'. But for him, instrumental interactions with 'theoretical objects' also provide thick epistemic access. So, Azzouni (2004, 384) suggests that he has a sufficient condition for commitment to the reality of at least *some* of the theoretical entities posited by a scientific theory:

> (I)f thick epistemic access meets the tracking requirement, then we can take the theoretical entities (that we have thick epistemic access to) to be real for the same reasons and on exactly the same grounds that we can take observational entities to be real.

What needs to be stressed is that both the theoretical irrealist and Azzouni take it that the tracking requirement normatively constrains access to the real. They disagree on exactly what judgements of ontic commitment this requirement licenses. *If* it turns out that there is a sense in which thick epistemic access to the real requires *thin* epistemic access to it, then Azzouni's attempt to contrast the two kinds of access collapses. This is what, I think, is the case.

Note that I don't want to deny that there *can* be thick epistemic access to the real. But I do want to stress that (a) Azzouni concedes far too much to the theoretical irrealist, because he accepts one of her main presuppositions (better: prejudices); and (b) Azzouni (as well as the theoretical irrealist) fail to see that thick epistemic access requires the *confirmation* of relevant theories.

## 5.2 On the epistemic authority of observation

Let's concentrate on observation, though what I shall say is also relevant to instrument-based thick access. Consider a claim that there is a chair next door. Relative to the evidence I now have, this claim is a hypothesis. It's a hypothesis about an observable entity, to be sure; but a hypothesis, nonetheless. So, I need to confirm it. Isn't the issue settled by just *looking* at the chair? And isn't this an exemplar of thick epistemic access? Isn't then otiose (or pedantic) to talk of *confirmation* is this case? Surely, one might say, the relevant epistemic process (i.e., looking) is sensitive to the presence and the properties of the chair and that's the end of the matter.

How did we get to the end of the matter? Observations rely on theory in two ways. One is the standard way: all observation is theory-laden. But neither Azzouni nor I take issue with this. The *other* way in which observations rely on theory has been made prominent by Sellars's (1956) attack on the Myth of the Given. This way is *very* relevant to the normative status of observation. Undoubtedly, observation is a main process by which we form beliefs, but an observation cannot have *epistemic authority* unless it can be evaluated as correct or incorrect. This evaluation will, inevitably, depend on theories of many sorts. They will be theories about why direct observation is, generally, a reliable way to find out facts about things like chairs and stones; theories about when an observer is a reliable indicator of her environment, and so on. Theories will not only tell when the observation has epistemic authority; they will also (if sufficiently developed) *correct* observations. As we move away from rudimentary observations of middle-sized objects, these theories will be more complex and demanding. They will be theories about how an instrument works, and why it is a reliable indicator of some things. More importantly, they will be theories about how what is observed using the instrument is correlated with (or is caused by) some properties of the unobservable entity that is detected by the instrument. It's irrelevant that these theories hardly ever function as *explicit* premises in our every-day undertakings. They are mostly internalised by the observers as they learn to use instruments, to calibrate them, to draw conclusions based on the data they collect using them and so on. Even so, they are pre-supposed for the evaluation of observation and hence for its epistemic authority.

Things turn out to be even more complicated. What are *really* observed in relatively sophisticated scientific contexts (e.g., in the context of an experiment) are not the phenomena themselves but *data*. This distinction between data and phenomena is an all-important one and has been rightly stressed in a seminal paper by Jim Bogen and James Woodward (1988). Strictly speaking, observations are of data (e.g., temperature readings, or clicks, or pointer readings).[2] The phenomena (e.g., the melting point of a substance or the path of a planet) are abstracted from the data by means of a number of sophisticated modelling techniques based on a rather substantive set of assumptions and theories. When a phenomenon is established (e.g., that lead melts at 327 degrees centigrade or that Mars describes – approximately – an ellipse) there is no strict observation of the phenomenon itself. Nor can it be said, in a straightforward manner, that the epistemic processes we use to establish (and come to know) this phenomenon track the relevant properties of, say, lead (or of Mars). The datum *327 degrees centigrade* might not even be within the data points that are collected by measurements of temperature.

Typically, what happens is the following. Scientists collect data and use methods such as controlling for possible confounding factors, data-reduction and statistical analysis to evaluate them. They make assumptions about, and try to check, the reliability of the equipment involved in the experiment. They make assumptions about the possible sources of systematic error and try to eliminate them (or correct the data in their light). They also make assumptions about the causes of variation in the data points they collect. Only when all this is in place, can there be a reliable *inference* from the data (or a subset of them) to the phenomenon. All this is explained in great detail by Bogen and Woodward

(1988). The relevant point here is that the establishment of the epistemic authority of what is normally called the observable phenomenon (e.g., the melting point of lead) is a rather complicated process which relies essentially on background theory. If all these theories somehow fail to be adequate (or well-confirmed, I should say), the observed phenomenon is called into question. This does not imply that before we establish, say, the melting point of lead, we need detailed theories of *why* lead melts at this point. This will typically be the product of further theoretical investigation. But it does imply that establishing the reliability (and hence the epistemic authority) of the data as a means to get to stable phenomena relies indispensably on some prior theories. Observation is not epistemically privileged *per se*. Its epistemic privilege is, in a certain sense, parasitic on the epistemic privilege of *some* theories.

Azzouni might have a reply to all this. In Azzouni (2000), he argues that the application and the empirical testing of theories relies on a body of 'gross regularities' that function *independently* of any scientific theory (in the sense that they cannot, at least in full, be appropriated by any scientific theory). These gross regularities have, as it were, a life of their own. Some of them are 'articulated', while others are 'unarticulated': they amount to rather detailed knowledge-how, which cannot even be expressed propositionally (cf. ibid., 63–8). Azzouni takes observation as well as instrument-based probing to rely heavily on such gross regularities. The most mundane case concerns the middle-sized objects (where the gross regularities concern e.g., the accessibility and the relatively stable behaviour of these objects). The most exciting case concerns viruses and sub-atomic particles (where the gross regularities concern e.g., patterns of behaviour of the particles, or connections between their properties and certain observable behaviours – cf. ibid., 116–17). So, he might well argue that observation (as well as instrument-based probing) is epistemically privileged precisely because it relies on such gross regularities that are, by and large, independent of theory.

Let me grant, for the sake of the argument, these gross regularities. It could be easily argued that Azzouni overstates his case for these gross regularities, but nothing hangs on this at present. Azzouni's appeal to these gross regularities creates more difficulties for his tracking requirement (as a *normative* constraint) than he thinks. The reason is this. Even if most of the theories mentioned a couple of paragraphs above as part of my claim that observation relies on theory for its normative import are taken to capture gross regularities (which they do not), these theories will involve ordinary and, sometimes, elementary *inductive* generalisations. The facts (regularities – gross or not) reported by these generalisations cannot be accessed in a thick way. No general fact can be accessed in a thick way (in Azzouni's sense). Commitment to these facts can only be the outcome of *thin* access to them: that our beliefs about them have been confirmed and that this gives us reasons to accept as true the generalisations that report them.[3]

The conclusion I want to draw will be obvious: there are no self-authenticating observational epistemic processes. If their epistemic authentication has to come from theories, all these theories have to be themselves authenticated, which, in this context at least, means *confirmed*. If confirmation, even if rudimentary, of all these theories is inescapable before we start taking seriously the idea of an *epistemically authoritative* thick access to the real, there is a sense in which *the very possibility of thick epistemic access presupposes (conceptually) thin epistemic access*. For, by hypothesis, confirmation gives us only thin access to the real. Thick access rests on (conceptually and epistemically) prior *thin* access, that is, it rests on commitment to whatever the relevant confirmed theories posit (including relevant regularities).

Let me summarise. It does not *really* matter whether we have thick or thin epistemic access to whatever is posited by our best theories of the world, provided that (a) what is posited has what it takes to be real and (b) our epistemic processes are such that we end up with correct beliefs about these posits, that is, beliefs that resonate with what these posits are and what properties they have. On some occasions, the theory itself (may be, in conjunction with other theories) may lead us to expect that a certain posit can be tracked by observations and instruments. But if the theory does *not* entail this for a posit, which is nonetheless explanatorily useful, this is *no* reason to suspect this posit. Our beliefs may track it not through observations and instruments, but in an alternative way.

## 5.3 The picture of the levels

What is this alternative way? I want to motivate a view that challenges the basic assumption shared by theoretical irrealists and Azzouni, namely, that commitment to observables is (should be) grounded on the epistemically privileged status of observation. I take my cues from Quine and Sellars. Despite their many differences, they both argued for two important theses. *First*, the theoretical-observational distinction is not ontological, but rather methodological. We are not talking here about two distinct senses of real, a theoretical existence and a non-theoretical one. There is just one sense of real at play. *Second*, the theoretical–observational distinction does not mark an ultimate epistemic difference either. It's not that observational claims can be known to be true, whereas theoretical claims cannot. It's not even that different, in principle, epistemic methods are employed for knowing them.

Quine's (1960, 22) master argument for the reality of molecules and the like is that 'they are on a par with the most ordinary physical objects'. As he explains:

> The positing of those extraordinary things (molecules and their extraordinary ilk) is just a vivid analogue of the positing or acknowledging of ordinary things: vivid in that the physicist audibly posits them for recognised reasons, whereas the hypothesis of ordinary things is shrouded in prehistory. (ibid.)

Epistemically speaking, molecules and chairs are also on a par, given that, for Quine, the evidence we have is couched in terms of 'surface irritations'. As Quine (1955, 250) puts it:

> If we have evidence for the existence of the bodies of common sense, we have it only in the way in which we may be said to have evidence for the existence of molecules. The positing of either sort of body is good science insofar merely as it helps us to formulate laws—laws whose ultimate evidence lies in the sense data of the past, and whose ultimate vindication lies in anticipation of sense data of the future.

Quine's move is to challenge the theoretical irrealist with a *tu quoque*: if you doubt the reality of molecules, you should doubt the reality of the bodies of common sense. Note that Quine does not need to appeal to something like Azzouni's thick epistemic access, for he takes the issue of how we know what there is to be subsumable under the issue of 'the evidence for truth about the world', the 'last arbiter' of which 'is so-called scientific method, however amorphous' (1960, 23). Sometimes, a certain process might directly track the properties of a real object, be it observable or unobservable (cf. 1955, 253). But this is a bonus, as it were. And it's not as if Quine thinks that thick epistemic access is already in place when we form beliefs about common sense bodies. A rudimentary version of scientific method rules there too. Quine (1981, 20) says explicitly that

> Even our primordial objects, bodies, are already theoretical (...). Whether we encounter the same apple the next time around, or only

> another one like it, is settled if at all by inference from a network of hypotheses that we have internalised little by little in the course of acquiring the non-observational superstructure of our language.

Sellars's approach is interestingly different from Quine's. On the one hand, Sellars (1956) resists the temptation that there are sense-data which can play an evidential role. On the other hand, he (1963) offers a master argument for commitment to the unobservable entities posited by scientific theories, namely, that they play an ineliminably explanatory role. In order to formulate this argument, he had to resist what he aptly called the *picture of the levels*. According to this, the realm of facts is layered. There is the bottom level of observable entities. Then, there is the intermediate level of the observational framework, which consists of empirical generalisations about observable entities. And finally, there is yet another (higher) level: the theoretical framework of scientific theories that posit unobservable entities and laws about them. It is part of this picture that while the observational framework is explanatory of observable entities, the theoretical framework is useful only for the explanation of the inductively established generalisations of the observational framework. But then, Sellars notes, the empiricist will rightly protest that the higher level is dispensable. For all the explanatory work vis-à-vis the bottom level is done by the observational framework and its inductive generalisations. Why then posit a higher level in the first place?

Sellars's diagnosis is that this picture is misleading. A Sellarsian realist should reject it. Sellars's argument is that the unobservables posited by a theory explain *directly* why (the individual) observable entities behave the way they do and obey the empirical laws they do (to the extent that they do obey such laws). So, he resists the idea that the theoretical framework has as its prime function to explain the empirical generalisations of the observational framework. As Sellars (1963, 121) graphically states it:

> Roughly, it is because a gas is—in some sense of 'is'—a cloud of molecules which are behaving in certain theoretically defined ways, that it obeys the empirical Boyle–Charles law.[4]

Sellars claimed that unobservable entities are indispensable because they also explain why observational generalisations are, occasionally, violated; why, that is, some observable entities do not behave the way they should, had their behaviour been governed by the observational generalisation. A good way to state Sellars's point is by noting that the observational generalisations will be either false or *ceteris paribus*. Think, for instance, of the generalisation that everything that is left unsupported falls to the ground. What about helium balloons, then? If these generalisations are false, the theory which is made of them will be *false* too, and hence will have to be replaced by a truer theory (one, for instance, which explains why most objects fall to the ground but helium balloons do not). If, on the other hand, these generalisations are *ceteris paribus* true, again it is theories of the theoretical framework (and their concomitant positing of unobservables) that explain the circumstances under which they hold strictly and the circumstances under which they fail to hold. In this case too, the theory that is made of *ceteris paribus* generalisations will have to be replaced by another truer theory that explains their limitations.[5]

Sellars (1963, 122) takes it that insofar as theories 'establish their character as indispensable elements of scientific explanation', they also establish themselves 'as knowledge about what *really* exists'. To a good approximation, what he has in mind is that, ultimately, scientific explanation proceeds via the *theoretical identifications* of observable entities with unobservables. The latter not only explain the observable behaviour of some observable entities; they really *are* the things which we thought of as independently existing observable entities. There isn't a table *and* a swarm of molecules. There is just a swarm of molecules. It's not puzzling, then, that we should be committed to unobservables. That's the only thing we *can* be committed to if we want to explain, and come to have *true* beliefs (and not just *ceteris paribus* observational generalisations) about, the entities which populate the observational framework.

An objection to this part of my argument might go as follows. My use of 'commitment' is ambiguous. For my argument to go through, commitment needs to be understood in a *de re* sense. But it is very doubtful that ontic commitment is a *de re* propositional attitude. It seems, rather, to be a *de dicto* attitude, which would nullify the argument made here, since substitutivity of identicals cannot be assumed. By way of reply, I should note that there is certainly a sense in which 'commitment' to an entity could be understood as a propositional attitude: when, for instance, we read '*S* is committed to *x*' as '*S* believes that there are *x*' (cf. van Fraassen 2000b, 1657–8). Yet, I am not sure it's *useful* to understand 'commitment' to an entity as a propositional attitude. And if it is not, I am no longer sure why commitment to an entity could not be *de re*. The relevant literature is vast and interesting. But there is a way for my argument to proceed that *seems* unobjectionable. I would follow Michael Jubien's (1972) formulation of ontic commitment: *T assumes x*, where *T* is a theory and *x* is *something else* (e.g., an entity) and the ontic commitment is a relation between a theory and this something else. There are quite a few problems with this proposal. But as Jubien (1972, 381–2) has shown, there is a clear case in which the substitutivity of identicals *can* be assumed, so long as the theory *itself* provides the ground for the substitution. Suppose that *T* assumes *a*. If $a = b$ is part of *T* (e.g., a theorem of *T*), then it *can* be inferred that T assumes *b*. It should be clear that this fits Sellars's proposal (and my argument) perfectly. For, the required theoretical identifications are part of the theory. To press the example above, since it is the theory itself that identifies tables with swarms of particles, if the theory assumes tables, it also assumes swarms of particles.

Perhaps, we start, unreflectively, with the idea that the observational framework consists of a domain of *sui generis* objects, to which we have perceptual access. Then we come to realise that we have been misled by the picture of the levels. We come to accept that the so-called observable entities are, in some sense, 'constructs'. From all this, it does *not* follow that we cannot have observational access to the so-called theoretical entities. Quite the contrary. If the theoretical identifications hold, such access is guaranteed. Indeed, Sellars's attack on the myth of the given was based, partly, on his claim that theoretical concepts and theoretical terms can gain a 'reporting role'. Nor, of course, does it follow that all so-called theoretical entities will be identified with observables. Still, their indispensable role in explaining the selected phenomena of which the theory is a theory of is enough to make us say that they have what it takes to be real.

There is certainly a sense in which Sellars can honour Azzouni's call for thick access the real. He insists that when one learns a scientific theory, one learns to 'tell by looking' that certain theoretical states of affairs obtain (Sellars 1977, 183). He takes it to be the case that 'a theoretical framework can achieve first-class status only if a proper subset of its expressions acquire a direct role in observation' (ibid.) But this is a methodological criterion which links the theoretical framework to intersubjective perceptual responses. There is also a sense in which the Sellarsian approach shows the limits of thick access. To say the least, there may be further theoretical identifications between different theoretical entities, which are ontically committing, because relevantly explanatory, without directly satisfying the tracking requirement.

What follows from the Sellarsian approach is that good theories go hand in hand with existence claims and knowledge of them.[6] 'Good theories', as Sellars (1963, 117–18) suggests, are those theories that explain directly the behaviour of observable entities by positing unobservables. What, for Sellars, makes these theories *good* is that they license an inference from some observational premises to a theoretical conclusion, for example, that molecules exist. This inference is certainly an explanatory inference; something like an *inference to the best explanation*. Sellars's important spin on this inference is that he weds it to the denial of the picture of the levels: either we have no explanation of what we take to be observable entities and their behaviour (i.e., either we just have a *ceteris paribus* inductive systematisation of them) or we have to accept that there are unobservables.

I prefer Sellars's account to Quine's for two reasons. First, Sellars's yields a more robust realism than Quine's, especially if we see Quinean realism in the light of his overall philosophical views concerning the immanence of truth, ontological relativity and the like. Second, Sellars's account makes more explicit than Quine's that ontic commitment and explanatory indispensability go hand in hand. But we shouldn't lose sight of their important connections. They both urge us to accept the reality of unobservable entities based on the role that these unobservables play in explaining observable entities and phenomena.

Following Sellars (and Quine), we have an alternative picture of how the real is tracked. Put in a slogan form: the real is tracked via good (in Sellars's sense) theories. In fact, we seem to have a better picture than Azzouni's. We now have a way to argue that what is posited has what it takes to be real. The Sellarsian argument can be seen as supplying an independent argument to support the claim that what our epistemic processes track *is* the real. Besides, the Sellarsian argument reveals the root problem with theoretical irrealism: it is captive of the picture of the levels. The Sellarsian argument is precisely that the theoretical irrealist is left with no ontic commitments at all, if she eschews commitments to unobservables. The supposed observational framework is a *false* theory. Not only is there nothing deeper (like sense contents) for the observational framework to be explanatory of, but also 'there *really* are no such things as the physical objects and processes of the common sense framework' (Sellars 1977, 174).[7] It's also interesting that the Sellarsian argument can even accommodate Azzouni's call for thick epistemic access – but as a *bonus*, or, perhaps, as a *double check*.

## 5.4 Quinean virtues

Azzouni rightly worries about the epistemic credentials of the Quinean virtues. He discusses five of the virtues that Quine has put forward: simplicity, familiarity, scope, fecundity and success under testing. It is certainly legitimate to wonder whence these virtues acquire their supposed epistemic force. It is equally legitimate to argue that since confirmation depends on them, if they are found lacking in epistemic force, confirmation will not carry much epistemic weight either. I am quite willing to grant that these virtues enter the Quinean picture as a *deus ex machina*. This, of course, is not to imply that they lack epistemic force. Success under testing surely has some such force. Consider how Quine (1955, 247) introduces 'familiarity' as a virtue: 'the already familiar laws of motion are made to serve where independent laws would otherwise have been needed'. It's not hard to see that a way to read this virtue is that familiarity increases the degree of confirmation of a theory, since the so-called familiar laws are already well-confirmed ones. Or consider 'fecundity': 'successful further extensions of the theory are expedited'. Here again, the connection with an increase in the degree of confirmation is not far away. But to say that a factor $x$ has epistemic force for a belief (or theory) $y$ is to say that $x$ raises the probability of $y$'s being true. So, there is a sense in which the Quinean virtues have epistemic force. This last thought might remove the sting from Azzouni's (2004, 378) claim (a joke?) that the list of virtues might be augmented to include, for instance, 'crass political manipulation'. The latter does not have, even *prima facie*, epistemic force in the sense explained above.

The point remains, however, that the Quinean virtues need defence. This point is equally forceful against the Sellarsian view I have been defending, since, at least *prima facie*, talk of explanatory theories and talk of virtuous theories go hand in hand. But there is a sense in which the Sellarsian view can escape Azzouni's pressure on the virtues. To say of a theory $T$ that it offers an explanation of certain observable entities, of their lawlike behaviour and of their deviations from this, is to tell a story in terms of the theory's unobservable posits and their properties, as to what these observable entities are and how their lawlike behaviour comes about. This (causal-nomological) story is a good theory irrespective of the virtues that it might have. Differently put, it is the positing of the molecules (with their properties and their causal role) that does the explaining and *not* the virtues of the theory of molecules. The virtues can and do enter the picture in an important but indirect way: as parts of a *theory* of how the first-order explanatory stories told by scientific theories are (should be) appraised. Rock-bottom, as it were, is empirical success, and especially what may be called novel empirical success, that is, confirmed novel predictions. But it may happen that this is not enough to single out one first-order story, because more than one tie with respect to empirical success. It is for this reason that we need to introduce a *theory* of theory-appraisal, whose elements are the virtues. If we see the problem in this light, it is open to us to treat this theory for what it is: a *theory*. We then need evidence, that is empirical evidence, for the virtues that will go into the theory. This evidence can only come from the past record of first-order explanatory stories. Put it succinctly, the question we should try to answer is this: what kinds of virtues were possessed by scientific theories that were successful (in the strict sense of also yielding novel predictions) as explanations of certain observable entities, of their lawlike behaviour and of their deviations from this? It hardly needs stressing that we are far from having such a theory. But, I think, such a theory can legitimise the virtues.[8]

Incidentally, I am in broad sympathy with Azzouni's critique of confirmational *holism* (and of its concomitant *holistic realism*, if there is such thing). I think, to be sure, that confirmational holism makes coherentism much more plausible as a theory of justification. For, it engenders and guarantees some friction with the world, and hence it renders justification not a simple function of explanatory coherence and the Quinean virtues. But in so doing, it can lead (and has led) to excesses. It's absurd, I think, to say that when a theory is put to the test the whole science is put to the test. It may well be problematic to say that when a theory is put to the test the laws of logic and the mathematics are also tested. But an important insight of confirmational holism vis-à-vis theoretical assertions is that they are confirmable no less than (*prima facie*) observational ones. As I have argued elsewhere (cf. 1999, 125–7), the insight that confirmation can go all the way up to the higher levels of theory does not entail that *all* theoretical assertions are confirmed, and to the same degree.

Azzouni has tried to challenge this last conclusion. To be sure, he turns his fire against Kitcher's relevant views. But both Kitcher (1993a) and I (Psillos 1999) have argued for two theses. *First,* there is a plausible explanatory argument (the no-miracles argument) that takes realists from the fact that theories are successful to the claim that they are approximately true. *Second,* in buying into this argument, realists need not accept that all theoretical constituents of a theory are confirmed and can be deemed (approximately) true. Kitcher and I draw the line between the 'good' and the 'bad' parts of successful theories differently, but we both agree that confirmation is selective and that the theoretical constituents that are confirmed are those that essentially contributed to the successes of a theory.[9] Azzouni has two arguments against this line of thought, both of which are well known.

The first is this: 'A set of falsehoods, however, can have true implications (implication isn't falsity-preserving). If, therefore, there are scientific theories composed entirely of falsehoods, but with true (empirically exploitable) implications, then, at least in these cases, attempts to divide the true from the false this way will fail'. True, but irrelevant. As has already been noted in previous chapters, realists adopt a rigorous notion of empirical success, which includes *novel* predictions. Given this, it is very unlikely, realists argue, that all of the theoretical premises that are employed in the derivation of novel predictions are false. So, Azzouni (or anyone else for that matter) should show that there are such theories that cannot be divided into true and false parts.

The second of Azzouni's arguments is this. In typical cases, 'successful theories are evaluated against a background of possible theories where ontological possibilities are rich; and so there is good reason to distrust the inferences from the success of these theories (and the instruments these theories are about) to their truth' (Azzouni 2004, 382). And he adds: '"success to truth" inferences only make sense when the ontological options excluded are meagre' (ibid., 382). It's not hard to see that Azzouni's point is none other than the old chestnut of the underdetermination of theories by evidence. I don't intend to repeat here the usual strategies that realists follow to block this argument (see Chapter 1, Section 1.2 and Psillos 1999, Chapter 8). My reply, instead, is minimal. Azzouni employs the foregoing argument in order to promote his own thick epistemic access. Presumably, in ordinary observation (where observable entities are involved), as well as in (some) instrument-based tracking, the ontological options are not rich. But who is to decide this? Relative to a sceptical scenario, the ontological options are wildly rich, even when it comes to ordinary observation.[10] Azzouni is surely right when he says that in typical cases of observations (of observables) 'we are severely constrained in the ontological options it's reasonable to even consider' (Azzouni 2004, 381). But the emphasis in this claim should surely be on the word 'reasonable'. If, as indeed happens, what options are reasonable to consider when it comes to observables depends on several background theories and assumptions (e.g., that there must be some physical cause of the extinction of the dinosaurs), it is surely open to someone to argue that analogous background theories and assumptions make it reasonable to restrict the ontological options when it comes to theoretical inferences from success to truth. In any case, the way Azzouni describes things does not show that theoretical inference from success to truth is unjustified or irrational. It only shows what we already knew, namely, that it is far more risky.

# 6
# Cartwright's Realist Toil: From Entities to Capacities

Cartwright has been both an empiricist and a realist. Where many philosophers thought that these two positions are incompatible (or, at any rate, very strange bed-fellows), right from her first book, the much-discussed and controversial *How the Laws of Physics Lie*, Cartwright tried to make a case for the following view: if empiricism allows a certain type of method (*inference to the most likely cause*), an empiricist cannot but be a scientific realist – in the metaphysically interesting sense of being ontically committed to the existence of unobservable entities. Many empiricists thought that since empiricism has been traditionally anti-metaphysics, it has to be anti-realist. One of the major contributions that Cartwright made to the philosophy of science is precisely this: there is a sense in which metaphysics can be respectable to empiricists. Hence, scientific realism cannot be dismissed on the grounds that it ventures into metaphysics. To be sure, the metaphysics that Cartwright is fond of is not of the standard a priori (or armchair) sort. It is tied to scientific practice and aims to recover basic elements of this practice (e.g., causal inference). But it is metaphysics, nonetheless.

Cartwright's realism has been described as 'entity realism'. This is not accidental. She repeatedly made claims such as 'I believe in theoretical entities' (1983, 89 – see also 92). Typically, she contrasted her commitment to entities to her denial of 'theoretical laws'. In Sections 6.1 and 6.2, I shall examine in some detail the grounds on which Cartwright tried to draw a line between being committed to entities and being committed to theoretical laws, and I shall find them wanting. Section 6.2 will also claim that the method Cartwright articulated, *inference to the most likely cause*, is important but incomplete. Specifically, I shall claim that there is a more exciting method that Cartwright herself describes as *inference to the best cause* (IBC), which, however, is an instance, or a species of *inference to the best explanation.* But Cartwright has been against inference to the best explanation (IBE). So, Section 6.3 will consider and try to challenge Cartwright's central argument against IBE.

At least *part* of the motivation for her early, restricted, realism was a certain understanding of what scientific realism is. She took scientific realism to entail the view that the world has a certain hierarchical structure, where the more fundamental laws explain the less fundamental ones as well as the particular matters of fact. In *The Dappled World,* she rightly disentangled these issues. 'Nowadays', she says, 'I think I was deluded by the enemy: it is not *realism* but *fundamentalism* that we need to combat' (Cartwright 1999, 23). What emerges quite clearly from her later writings is that Cartwright does *not* object to realism. Rather, she objects to Humeanism about laws, causation and explanation. Insofar as Humeanism is a metaphysics independent of scientific realism, Cartwright is a more full-blown realist, without being Humean. In Section 6.4, I shall discuss in some detail Cartwright's central non-Humean concept, namely, *capacities.* Cartwright is a strong realist about capacities. They are the fundamental building blocks of her metaphysics. But there seem to be a number of problems with capacities. Though we can easily see how attractive it is to be a realist about capacities, it's really hard to be one. Humeanism is certainly independent of scientific realism, but as I shall argue we have not been given compelling reasons for a non-Humean metaphysics of capacities.

## 6.1 Causal explanation

One of Cartwright's (1983) central claims is that causal explanation is ontically committing to the entities that do the explaining. Here are some typical statements of it:

> That kind of explanation succeeds only if the process described actually occurs. To the extent that we find the causal explanation acceptable, we must believe in the causes described (1983, 5).
>
> In causal explanations truth is essential to explanatory success (1983, 10).
>
> But causal explanations have truth built into them (1983, 91).
>
> (...) existence is an internal characteristic of causal explanation (1983, 93).

These assertions are not all equivalent with one another, but I will not dwell on it. There is indeed something special with causal explanation. Let's try to find out what it is. As a start, note that it is one thing to say that causal explanation is ontically committing, but it is quite another thing to say what a causal explanation *is*. Let's take them in turn.

### 6.1.1 Ontic commitment

If *c* caused *e*, there must be events *c* and *e* that are thus causally connected. Causation is not quite the same as causal explanation, but causes do explain their effects and there is, to say the least, no harm in saying that if *c* causes *e* then *c* causally explains *e*. This feature of causal explanation by virtue of which it is ontically committing to whatever does the causing is not peculiar to it. Compare the relation *c preceded e*: *c* must exist in order to precede *e*. So, Cartwright's claim is an instance of the point that the relata of an actual relation *R* must exist in order for them to be related to each other by *R*. I think this is what Cartwright should mean when she says that '(...) existence is an internal characteristic of causal explanation' (1983, 93).

An equivalent way to show that causal explanation is ontically committing is this. To say that the statement '*c* causally explains *e*' is ontically committing to *c* and *e* is to say that '*c* causally explains *e*' is *true*. This way of putting things might raise the spectre of van Fraassen, as Hitchcock (1992) reminds us. Couldn't one just accept that '*c* causally explains *e*' without believing that it is true? If so, couldn't one simply *avoid* the relevant ontic commitments to whatever entities are necessary to make this statement true? Insofar we can make sense of an attitude towards a statement with a truth-condition which involves acceptance but not belief, van Fraassen is on safe ground here. He is not forced to believe in the truth of statements of the form '*c* causally explains *e*'. Cartwright's point, however, is not meant to be epistemic. Her point is two-fold. On the one hand, she stresses that we cannot avoid commitment to the things that are required to make our assertions true. On the other hand (and more importantly), insofar as we do make *some* assertions of the form '*c* causally explains *e*' (e.g., about observable events such as short-circuits and fires, or aspirins and headaches), there is no reason not to make others (e.g., about unobservable entities and their properties).

Causal explanation is egalitarian: it sees through the observable–unobservable distinction. It is equally ontically committing to both types of entity, precisely because the relation of causal explanation is insensitive to the observability of its relata. What matters for ontic commitment is the causal bonding of the relata of a causal explanation. Cartwright's point is that there is just one way to be committed to entities (either observable or unobservable) and is effected through causal explanation.

### 6.1.2 What exactly is a causal explanation?

This remains an unsettled question, even after it is accepted that causal explanation is ontically committing. The question, in a different form, is this: what exactly is the relation between *c* and *e* if *c* causally explains *e*? In the literature, there have been a number of attempts to explain this relation. I am not going to discuss them here.[1] Cartwright has offered a gloss of the relation *c causally explains e*. Cartwright (1983, 25–6), put forward an early version of the contextual unanimity principle, namely, the idea that *c* causes *e* iff *c* increases the probability of *e* in all situations (contexts) which are causally homogeneous with respect to the effect *e*. I will not dwell on this principle here. But one thing is relevant. Although principles such as the above do cast some light on the notion of causal explanation, they do not offer an analysis of it, since they presuppose some notion of causal law, or some notion of causally homogeneous situation. Cartwright is very clear on this when she says, for instance, that what makes the decay of uranium 'count as a good explanation for the clicks in the Geiger counter' is not the probabilistic relations that obtain between the two events, 'but rather the causal law – "Uranium causes radioactivity"' (1983, 27).

Still, it might be said that though Cartwright does not offer 'a model of causal explanation' (1983, 29) she does constrain this notion by objecting to certain features that causal explanation is taken to have. Most centrally, she objects to the deductive-nomological model of causal explanation. But it is not clear, for instance, that she takes a *singularist* account of causal explanation. It seems that she doesn't. For she allows that certain 'detailed causal principles and concrete phenomenological laws' are involved in causal explanation (ibid., 8). Her objection is about laws captured by 'the abstract equations of a fundamental theory' (ibid.) Even if she objects to the thesis that *all* causal explanation should be nomological, she doesn't seem to object to the weaker thesis that *at least some* causal explanation should be nomological. In any case, it's one thing to deny that the laws involved in causal explanation are the abstract high-level laws of a theory and it is quite another to deny that laws, albeit low-level ones, are involved in, or ground, causal explanation. For all I know, Cartwright (1983) does *not* deny the latter.

Here is the rub, then. If *laws* are presupposed for causal explanation, it's no longer obvious that in offering causal explanations we are committed *just* to the relata of the causal explanation. To say the least, we should also be committed to a Davidson-style compromise that *there are* laws that govern the causal linkage between cause and effect. These laws might not be stateable or known, but they cannot be eliminated. This is not the end of the story. Considering Davidson's idea, Carl Hempel noted that when the existence of the law is asserted but the law is not explicitly stated, the causal explanation is comparable to having 'a note saying that there is a treasure hidden somewhere' (Hempel 1965, 349). Such a note would be worthless, unless 'the location of the treasure is more narrowly circumscribed'. Think of it as an advice: where there is causal explanation, search for the law that makes it possible. It's a side-issue whether this law will be a fundamental one or a phenomenological one or what have you. This is a worry about the kinds of law there are and not about the *role* of laws in causal explanation.

My first conclusion: Cartwright's advertised entity realism underplays her important argument for ontic commitment. In offering causal explanations we are committed to more than entities. We are also committed to laws, unless of course there is a cogent and general story to be told about causal explanation that does *not* involve laws. Note that it is not a reply to my charge that there might be singular causal explanation. This is accepted by almost everybody – given the right gloss on what it consists in. Nor would it be a reply to my charge that, occasionally, we do not rely on laws to offer a causal explanation. A suitable reply would have to show that causal explanation is fully disconnected from laws. This kind of reply might be seen as being offered by Cartwright when she introduces capacities. But, as we shall see in Section 6.4, it is questionable that we can make sense of capacities without reference to laws.

## 6.2 Causal inference

Given the centrality of causal explanation in Cartwright's argument for realism, one would have expected her to stay firmly in the business of explaining its nature. But Cartwright does something *prima facie* puzzling. She spends most of her work (Cartwright 1983) in an attempt to cast light on the nature of the *inference* that takes place when a causal explanation is offered and on the conditions under which this inference is legitimate. (Doesn't that remind you of someone else? Right guess!) One way to read what Cartwright does is this: she is concerned with showing when a potential causal explanation can be accepted as the actual one. More specifically, she intends to show that there is something special in causal inference that makes it sound. She (ibid., 6) says:

> Causal reasoning provides good grounds for our beliefs in theoretical entities. Given our general knowledge about what kinds of conditions and happenings are possible in the circumstances, we reason backwards from the detailed structure of the effects to exactly what characteristics the causes must have to bring them about.

Thus put, causal reasoning is just a species of ampliative reasoning. From an epistemic point of view, that the explanation offered in this reasoning is *causal* (i.e., that it talks about the putative causes of the effects) is of no special importance. What matters is what *reason* we have to accept the conclusion about the putative cause.

Cartwright explicitly draws a contrast between 'theoretical explanation' and 'causal explanation' (ibid., 12). But this is, at least partly, unfortunate. It obscures the basic issue at stake. *Qua* inferential procedures, causal explanation and theoretical explanation are on a par. They are species of ampliative reasoning and the very same justificatory problems apply to both of them (perhaps to a different degree).

Cartwright does think that there is something special in the claim that the inference she has in mind relies on a *causal* explanation. She calls this inferential process 'inference to the most likely cause' (ibid., 6) – henceforth, IMLC. But there is a sense in which the weight is on the 'most likely' and not on the 'cause'. It's just that Cartwright thinks that it's most likely to get things right if you are looking for causes than if you are looking for something else (e.g., general theoretical explanations). Before we see whether this is really so, let us press the present point a bit more.

### 6.2.1 Inference to the most likely cause

What kind of inference is IMLC? An obvious thought is that we infer the conclusion (viz., that the cause is *c*) if and only if the probability of this conclusion is high. But this is a general constraint on *any* kind of ampliative inference with rules of detachment; hence there is nothing special in IMLC in this respect. A further thought might be that in the case of IMLC there is a rather safe way to get the required high probability. The safety comes from relevant background knowledge of all sorts: that the effect has a cause, because in general effects do have causes; that we are offered a rather detailed story as to what the causal mechanism is and how it operates to bring about the effect; that we have controlled for all(?) other potential causes etc. (cf. 1983, 6). All this is very instructive. However, thus described, IMLC gets its authority not as a special mode of inference where the weight is carried by the claim that *c causally explains e*, but from whatever considerations help increase our confidence that the chosen hypothesis (viz., that it was *c* that caused *e*) is likely to be true. If these considerations are found wanting (if, for instance, our relevant background knowledge is not secure enough, or if we do not eliminate all potential alternative causes, or if the situation is very complex) the claim that *c* causally explains *e* is inferentially insecure. It simply cannot be drawn, because it is not licensed as likely.

My present complaint can be strengthened. Consider what Cartwright (1983, 82) says: '(...) causal accounts have an independent test of their truth: we can perform controlled experiments to find out if our causal stories are right or wrong'. If we take this seriously, then all the excitement of IMLC is either lost or becomes parasitic on the excitement of a controlled experiment. It is lost if for every instance of an IMLC it is required that a controlled experiment is performed to check the conclusion of the inference *independently*. What if the excitement of IMLC becomes parasitic on the excitement of a controlled experiment? Controlled experiments are indeed exciting. But their excitement comes mostly from the fact that they are designed to draw safe causal conclusions, irrespective of whether there is on offer a causal *explanation* of the effect. When it is established by a clinical trial that drug *D* causes relief from symptom *S*, we may still be in the dark as to *how* and *why* this is effected, what the mechanisms are, what the detailed causal story is and the like. I thought that causal explanation – *qua* inference – is exciting not just because we can get conclusions that are likely to be correct, but also because we get an understanding of *how* and *why* the effect is produced. So far, we have got only (or mostly) the former. The hard question remains unaddressed: what is this (if anything) in virtue of which a causal explanation – *qua* an explanatory story – licenses the conclusion that it is likely to be correct? Put in more general terms, the hard problem is to find an *intrinsic* feature of causal explanation in virtue of which it has a claim to correctness and not just an *extrinsic* feature, namely, that there are independent reasons to think it is likely.

### 6.2.2 Inference to the best cause

Cartwright seems aware of the need for such an intrinsic feature. Occasionally, she describes IMLC as 'inference to the best cause' (Cartwright 1983, 85). This is not just a slip. Reference to 'best cause' is not meant to *contrast* IBC to inference to the best explanation (IBE), by replacing 'explanation' with 'cause'. It is meant, rightly I think, to *connect* IBC to IBE. It is meant to base the inference (the detachment of the conclusion) on certain features of the connection between the premises and the conclusion, namely, that there is a genuinely explanatory relation between the explanation offered and the explanandum. The 'best cause' is not *just* a likely cause; it is a putative cause that explains the effect in the sense that it offers genuine understanding of how and why the effect was brought about. Cartwright says of Perrin's 'best cause': 'we are entitled to (infer the existence of atoms) because we assume that causes make effects occur in just the way they do, via specific, concrete causal process' (ibid., 85). If all we were interested in were high probability, we wouldn't go for specific, concrete causal processes – for the more detail we put in, the more unlikely they become. The specific, concrete causal processes matter for understanding, not for probability.

The upshot is that if we conceive causal inference as inference to the best cause, it is no longer obvious that it is radically different from what has come to be known as inference to the best explanation. The leading idea behind IBE – no matter how it is formulated in detail – is that explanatory considerations guide inference. The inference we are concerned with is ampliative – hence deductively invalid. But this is no real charge: inferential legitimacy is not solely the privilege of deductive inference. IBC can then be seen as a species of IBE. It's a species of a genus, whose *differentia* is that in IBC the explanations are causal.

If we think of causal inference as an inference to the most likely cause, the inferential weight is carried by the likeliness of the proposed causal explanation. So, it's not that a causal *explanation* is offered that licenses the inference. Rather, it is that this proposed explanation has been rendered likely. This rendering is *extrinsic* to the explanatory quality of the proposed explanation and relates to what we have done to exclude other potential explanations. On the other hand, if we think of causal inference as inference to the best cause, we are committed to the view that the inferential weight is carried by the explanatory quality of the causal explanation offered, on its own *and* in relation to competing alternatives.

Cartwright speaks freely of 'causal accounts' or 'causal stories' offered by causal explanations. The issue then is not *just* to accept that there must be entities which make these causal accounts true. It is also to assess these accounts *qua* explanatory stories. If we take IBC seriously, there must be ways to assess these accounts, and these ways must guide whether we should accept them as true. It seems then that we need to take account of explanatory virtues (a) if we want to make IBC have a claim to truth; and (b) if we want to tie this claim to truth not just to extrinsic features of causal explanation (e.g., that it is more likely than other potential explanations) but also to intrinsic features of the specific causal explanatory story.

Let me draw the conclusion of this section. Thinking of causal explanation as an inference to the best cause will require assessing the causal story offered and this is bound to be based on explanatory considerations that align IBC to IBE.

## 6.3 Why deny inference to the best explanation?

Cartwright (1983) resists IBE. She thinks she is not committed to IBE, when she vouches for IBC. So the issue is by no means over. Cartwright (1983, 4) explicitly denies that 'explanation is a guide to truth'. She discusses this issue quite extensively in Cartwright (1983). Here, I will focus on one of her arguments, which seems to me to be the most central: the argument from the falsity of laws. Before I go into this, allow me to note an interesting tension in her current views on the matter.

### 6.3.1 The transcendental argument

Cartwright has always resisted global applications of IBE. In particular, in Cartwright (1983), she tried to resist versions of the 'no-miracles argument' for realism. Consider her claim:

> I think we should instead focus on the causal roles which the theory gives to these strange objects: exactly how are they supposed to bring about the effects which are attributed to them, and exactly how good is our evidence that they do so? The general success of the theory at producing accurate predictions, or at unifying what before had been disparate, is of no help here. (ibid., 8)

The last sentence of this quotation is certainly overstated. But let's not worry about this now. In her current views, the general anti-theory tone in her work (Cartwright 1983) has been superseded by a more considered judgement about theories and truth. She concedes that 'the impressive empirical successes of our best physics theories may argue for the truth of these theories', but she denies that it argues 'for their universality' (Cartwright 1999, 4). Her talk about 'different kinds of knowledge in a variety of different domains across a range of highly differentiated situations' implies that *truth* is in the vicinity. For knowledge without truth is an oxymoron. So, her objections to IBE no longer aim to challenge the very possibility of a link between explanation and truth. Rather, they aim to block gross and global applications of IBE.

Let us look at Cartwright's (1999, 23) argument for 'local realism', which, as she says, is supposed to be a Kantian transcendental argument. The way she sets it up is this: We have X – 'the possibility of planning, prediction, manipulation, control and policy setting'. But without Φ – 'the objectivity of local knowledge' – X would be impossible, or inconceivable. Hence Φ. It's fully understandable why Cartwright attempts to offer a *transcendental* argument. These arguments are dressed up as deductive. Hence, they are taken not to have a problematic logical form. They compare favourably with IBE. But apart from general worries about the nature and power of transcendental arguments,[2] there is a more specific worry: *is the above argument really deductive?*

A cursory look at it suggests that it is: 'Φ is necessary for X; X; Therefore, Φ'. But it is misleading to cast the argument as above, simply because it is misleading to say that Cartwright's Φ is necessary for X. Kant thought that Euclidean geometry was necessary for experience. Of course, it isn't. He could instead have argued that *some* form of spatial framework is necessary for experience. This might well be true. But now it no longer deductively follows that Euclidean geometry must be true. In a similar fashion, all that Cartwright's argument could show is that something – call it Φ – is necessary for 'the possibility of planning, prediction, manipulation, control and policy setting'. But now, it no longer follows deductively that this Φ *must* be the realist's 'objective local knowledge', no matter how locally or thinly we interpret this. To say the least, this Φ could be just empirically adequate beliefs, or unrefuted beliefs, or beliefs that the world co-operates only when we actually try to set plans, make observations, manipulate causes and so on. Put in a different way, all that follows from Cartwright's transcendental argument is a *disjunction*: either objective local knowledge, or empirically adequate beliefs, or…is necessary for the possibility of planning, prediction, manipulation, control and policy setting. But which disjunct is true? Further argument is needed. There cannot be a transcendental deduction of objective local knowledge.

My suggestion is this: the move from the 'the possibility of planning, prediction, manipulation, control and policy setting' to a realist understanding of what needs to be the case for all of them to be possible can only be based on an inference to the best explanation: 'The objectivity of local knowledge' (as opposed to any other candidate) should be accepted on the grounds that it *best explains* 'the possibility of planning, prediction, manipulation, control and policy setting'. The moral then is that Cartwright's recent, more robust, realism can only be based on the very method that she has taken pains to disarm. We can now move on to look at the credentials of one of her stronger early arguments against IBE, namely, the alleged falsity of laws.

### 6.3.2 False laws?

In Cartwright (1983), one of the main theses is that explanation and truth pull apart. When laws come into the picture, this thesis seems to be the outcome of a certain *failure* of laws. She puts it like this:

> For the fundamental laws of physics do not describe true facts about reality. Rendered as descriptions of facts, they are false; amended to be true, they lose their fundamental, explanatory power. (ibid., 54)

We are invited to see that if laws explain, they are not true and if they are true, they do not explain. What Cartwright has in mind, of course, is what she calls fundamental or abstract laws as well as the covering-law model of explanation. If laws explain by 'covering' the facts to be explained, Cartwright says, the explanation offered will be false. If, she would go on, the laws are amended by using several *ceteris paribus* clauses, they become truer but do not 'cover' the facts anymore; hence, in either case, they do *not* explain the facts. The reason why covering laws do not explain has mostly to do with the fact that the actual phenomena are too complex to be covered by simple laws. Recall her example of a charged particle that moves under the influence of two forces: the force of gravity and Coulomb's force. Taken in isolation, none of the two laws (i.e., Newton's inverse-square law and Coulomb's law) can describe the actual motion of the charged particle. From this,

Cartwright concludes that each loses either its truth or its explanatory power. Here is her argument:

> The effect that occurs is not an effect dictated by any one of the two laws separately. In order to be true in the composite case, the law must describe one effect (the effect that actually happens); but to be explanatory, it must describe another. There is a trade-off here between truth and explanatory power. (ibid., 59)

I fail to see how all this follows. For one, it does *not* follow that there is not (worse: there cannot be) a complex law that governs the motion of massive *and* charged particles. If we keep our eyes not on epistemology (can this law be known or stated?) but on metaphysics (can there be such a law?), the above argument is inconclusive. For another, in the composite case, there is no formal tension between truth and explanation. In the composite case, none of the two laws (Newton's and Coulomb's) is strictly true of, in the sense of 'covering', the effect that actually happens. Why should we expect each of them *on its own* to 'cover' the complex effect? After all, the complex phenomenon is *governed* by both of them jointly, and hence it cannot be covered by each of them separately. This does not imply that laws lose their explanatory power. They *still* explain how the particle would behave if it was just massive and not charged or if it was charged but not massive. And they still contribute to the full explanation of the complex effect (i.e., of the motion of the charged *and* massive particle). To demand of each of them to be explanatory in the sense that each of them should somehow *cover* the actual complex effect is to demand of them something they cannot do. The laws do not thereby cease to be true, nor explanatory. Nor does it follow that they don't jointly *govern* the complex effect. *Governing* should not be conflated with *covering*.[3]

My argument so far might be inconclusive. But there is an important *independent* reason why we should take laws seriously. Laws individuate properties: properties are what they are because of the laws they participate in. Cartwright says: 'What I invoke in completing such an explanation are not fundamental laws of nature, but rather properties of electrons and positrons, and highly complex, highly specific claims about just what behaviour they lead to in just this situation' (ibid., 92). If it is the case that *no laws, no properties*, or if properties and laws are so intertwined that one cannot specify the former without the latter and conversely, some laws had better be true. For if they are not, then we cannot talk of properties either.[4]

This last point, however, is controversial, especially as of late. It relies on a Humean understanding of properties. And Cartwright is a non-Humean, more or less about everything. This observation is crucial. For it is Humeanism that is the real opponent of Cartwright's. Her capacities are non-Humean tendencies: causal powers. They are irreducible, primary and causally active constituents of the world. Similarly, her properties are non-Humean properties: they are active causal agents, which are identified via their causal role and their powers. So, it is not laws that determine what they are; rather, it is properties (capacities etc.) that determine what, if any, laws hold in the world. With all this in mind, let us turn our attention to her views about capacities. This is just one of her non-Humean themes. But it is the most central one.

## 6.4 Capacities

Cartwright has devoted two books in the defence of the claim that capacities are prior to laws (1989; 1999). She challenges the Humean view that laws are exceptionless regularities, since, she says, there are no such things.[5] How then does it appear that there *are* regularities in nature, for example, that all planets move in ellipses?

### 6.4.1 Nomological machines

Cartwright does not deny that there can be pockets of regular behaviour in nature. But she claims that where there is regular behaviour in nature, there is a *nomological machine* that makes it possible. A 'nomological machine' is

> a fixed (enough) arrangement of components, or factors, with stable (enough) capacities that in the right sort of stable (enough) environment will, with repeated operation, give rise to the kind of regular behaviour that we represent in our scientific laws. (Cartwright 1999, 50)

Nomological machines ensure that 'all other things are equal'. They secure the absence of factors, which, were they present, would block the manifestation of a regularity. Take Kepler's first law. This is not a strictly universal and unconditional law. Planets do (approximately) describe ellipses, if we neglect the gravitational pull that is exerted upon them by the other planets, as well as by other bodies in the universe. The proper formulation of the law, Cartwright argues, is: ceteris *paribus*, all planets move in ellipses. Now, suppose that the planetary system is a stable enough nomological machine. Suppose that as a matter of fact, the planetary system is (for all practical purposes) shielded: it is sufficiently isolated from other bodies in the universe, and the pull that the planets exert on each other is negligible. Under these circumstances, we can leave behind the *ceteris paribus* clause, and simply say that all planets move in ellipses. But the regularity holds only so long as the nomological machine backs it up. Were the nomological machine to fail, so would the regularity.

For the operation of a nomological machine, it is not enough to have a stable (and shielded) arrangement of components in place. It is not enough, for instance, to have the sun, the planets and the gravitational force in order for the planetary machine to run. Cartwright insists that it is the *capacities* that the components of the machine have that generate regular behaviour. For instance, 'a force has the capacity to change the state of motion of a massive body' (ibid., 51). Couldn't the nomological machine itself be taken to be a regularity? No, she answers: 'the point is that the fundamental facts about nature that ensure that regularities can obtain are not again themselves regularities. They are facts about what things can do' (1999, 4). But what exactly are capacities, that is, the things that things can do?

In Cartwright (1989, 9), she focused her attention on 'what capacities do and why we need them' and *not* on 'what capacities are'. What they are is the job of *The Dappled World*. Before, however, we examine what they are, let us see the main argument she offers as to why we need capacities.

### 6.4.2 Why do we need capacities?

#### *6.4.2.1 The Sellarsian argument*

As we saw in Chapter 5, Section 5.3, Sellars's (1963) master argument for commitment to the unobservable entities posited by scientific theories is that they play an ineliminable explanatory role. In order to formulate it, he had to resist the *picture of the levels*.

Cartwright offers an argument structurally similar to Sellars's in defence of capacities (cf. 1989, 163). She has in mind another possible layer-cake. The bottom-level is the non-modal level of occurrent regularities; the intermediate level is the level of (deterministic or statistical) Humean laws. The higher level is supposed to be a *sui generis* causal one.

*This* layer-cake, Cartwright notes, invites the thought (or the temptation) to do away with the higher-level altogether. All the explanatory work, it might be said, is done by Humean laws, endowed with modal force. The higher (causal) level could then be just seen as a higher *modal* level, with no claim to independent existence: it is just a way to talk about the intermediate level, and in particular a way to set constraints on laws in order to ensure that they have the required modal force. It is *this* layer-cake that Cartwright wants to resist. For her, the higher causal level is indispensable for the explanation of what regularities there are (if any) in the world. So, we seem to have a solid Sellarsian argument for capacities. But do we?

Before we proceed to examine this, an exegetical point is in order. Cartwright splits the higher (causal) level into two sub-levels: a lower sub-level of *causal laws* and a higher sub-level of *ascriptions of capacity*. She couches all this in terms of two levels of generality or more accurately of two levels of modality (ibid., 142). She says:

> (...) the concept of general *sui generis* causal truths—general causal truths not reducible to associations—separates naturally into two distinct concepts, one at a far higher level of generality than the other: at the lower level we have the concept of a causal law; at the higher, the concept of capacity. I speak of two levels of generality, but it would be more accurate to speak of levels of modality, and for all the conventional reasons: the claims at both levels are supposed to be universal in space and through time, they support counterfactuals, license inferences, and so forth. (ibid., 142)

Why do we need *two* causal levels? Why, in particular, do we need a level of *capacities*? To cut a long story short, Cartwright thinks that causal laws are kinds of causal generalisations relative to a particular population (cf. ibid., 144). They are causal, as opposed to Humean laws of association, mostly because the facts they report (e.g., that aspirins relieve headaches or that smoking causes cancer) cannot be fully captured by probabilistic relations among magnitudes or properties. Causal information is also required to specify the conditions under which they hold. A further thought then is that ascription of capacities is also necessary in order to remove the relativised-to-a-population character of causal laws. We don't just say that smoking causes cancer to population X. We want to say that smoking causes cancer, *simpliciter*. This claim (which is universal in character) is best seen as a claim about capacities: *C* causes *E*, means *C* carries the capacity *Q* to produce *E* (cf. ibid., 145). Capacities, then, are introduced to *explain* causal laws and to render them universal in character.[6]

What then of Cartwright's Sellarsian argument for capacities? I will focus on just one central problem. Sellars saves the higher level of electrons, protons etc. by focusing on the indispensable role this level plays in the explanation of singular observable phenomena or things. Similarly, one would demand of Cartwright's argument to show how capacities are indispensable for the explanation of occurrent regularities, without the intervening framework of Humean laws plus modal force. But it seems that there is a tension in her argument. Whereas in Sellars's case, the entities of the theoretical framework (unobservables) can be identified independently of the entities in the bottom framework, it is debatable that this can happen in Cartwright's case. Here there are conflicting intuitions. One is that we need regularities (or Humean laws) to identify what capacities things carry. Another (Cartwright's, I think) is that this is not the case. I am not entirely certain whose intuitions are right. But it seems to me that the Humean is on a better footing. Capacities might well be posited, but only *after* there has been a regular association between relevant event-types. No-one would mind ascribing to aspirin the capacity to relieve headaches, if that was the product (as indeed it is) of a regular association between taking aspirins and headaches' going away. 'Regular' here does not necessarily mean exceptionless. But, so much the better for positing capacities if the association happens to be exceptionless. To say the least, one could more easily explain how capacities have modal force. So, there is an important disanalogy between Sellars's argument for unobservables and Cartwright's argument for capacities, which casts doubt on the indispensability of positing capacities: in Cartwright's case, we need the lower level (regularities) to identify the entities of the higher level (capacities).

### *6.4.2.2 Single cases*

Cartwright insists that capacities might reveal themselves only occasionally or only in a single case. Consider what she says:

> 'Aspirins relieve headaches'. This does not say that aspirins always relieve headaches, or always do so if the rest of the world is arranged in a particularly felicitous way, or that they relieve headaches most of the time, or more often than not. Rather it says that aspirins have the capacity to relieve headaches, a relatively enduring and stable

> capacity that they carry with them from situation to situation; a capacity which may if circumstances are right reveal itself by producing a regularity, but which is just as surely seen in one *good* single case. The best sign that aspirins can relieve headaches is that on occasion some of them do. (Cartwright 1989, 3 – emphasis added)

This is surely puzzling. Just adding the adjective 'good' before the 'single case' does not help much. A 'good' controlled experiment might persuade the scientist that he has probably identified some causal agent. But surely, *commitment* to it follows only if the causal agent has a regular behaviour which can be probed in similar experiments. A single finding is no more compelling than a single sighting of a UFO. Single or occasional manifestations cast doubt on the claim that there is a stable and enduring capacity at play (cf. also Glennan 1997, 607–8).

Cartwright disagrees. In Cartwright (1999, 83), she advances what she calls the 'analytic method' in virtue of which capacity ascriptions are established. In Cartwright (2002, 435–6) she summarises her ideas thus:

> We commonly use the analytic method in science. We perform an experiment in 'ideal' conditions, *I*, to uncover the 'natural' effect *E* of some quantity, *Q*. We then suppose that *Q* will in some sense 'tend' or 'try' to produce the same effect in other very different kinds of circumstances. (...) This procedure is not justified by the regularity law we establish in the experiment, namely 'In $I$, $Q \rightarrow E$'; rather, to adopt the procedure is to commit oneself to the claim '*Q* has the capacity to *E*'.

What is the force of this claim? Note, first, that we don't have a clear idea of what it means to say that *Q* 'tends' or 'tries' to produce its effects. It seems that either *Q* does produce its effect or it doesn't (if, say, other factors intervene). Second, as Teller (2002, 718) notes, it is not clear how the 'trying' can be established by looking at a single case only. One thought here might be that if we have seen *Q* producing *its* effect at least one time, we can assume that it *can* produce it; and hence that it has the *capacity* to produce it. But I don't think this is the right way to view things. Consider the following three questions: (i) What exactly is *Q*'s effect? (ii) How can we know that it was *Q* which brought *E* about? And (iii) Wouldn't it be rather trivial to say that for each effect there is some capacity *X* which produces it? All three questions would be (more easily) answered if we took capacities to be regularly manifested. The 'regularity law' 'In $I$, $Q \rightarrow E$' makes the positing of a capacity legitimate. It is because (and insofar as) 'In $I$, $Q \rightarrow E$' holds that we can say that '$Q$ has the capacity to $E$', and not the other way around.

If the capacity $Q$ of $x$ to bring about $y$ was manifested regularly, one could say that the presence of the capacity could be tested. Hence, one could move on to legitimately attribute this capacity to $x$. But if a capacity can manifest itself in a *single* case, it is not clear at all how the presence of the capacity can be tested. Why, in other words, should we attribute to $x$ the capacity to bring about $y$, instead of claiming that the occurrence of $y$ was a matter of chance? So, there seems to be a tension between Cartwright's claim that capacities are manifestable even in single cases and her further claim that capacities are *testable*.[7]

So far, I have focused on the relation between capacities (the higher level) and regularities (the lower level). But there is also a problem concerning the two sub-levels of the higher level, namely, capacities and causal laws.[8] Do claims about the presence of capacities have *extra content* over the claims made by ordinary causal laws? So, do we really need to posit capacities? Take, for instance, the ordinary causal law that aspirin relieves headaches. If we ascribed to aspirin a *capacity* to relieve headaches, would we gain in content? There is a sense in which we would. Ordinary causal laws are *ceteris paribus*, whereas capacity claims are not. Since it is only under certain circumstances that aspirin relieves headaches, it is only *ceteris paribus* true that aspirin causes headache relief. But, Cartwright might say, once it is *established* that aspirin carries the *capacity* to relieve headaches, the *ceteris paribus* clause is removed: the capacity is *always* there, even if there may be contravening factors that block, on occasion, its manifestation. The problem with this attempt to introduce capacities is that the strictly universal character of claims about capacities *cannot* be established. If it is allowed that claims about the presence of capacities might be based on single manifestations, it is not quite clear what kind of *inference* is involved in the movement from a single manifestation to the presence of the capacity. Surely, it cannot be an inference based on any kind of ordinary inductive argument.[9] If, on the other hand, it is said that claims about capacities are established by ordinary inductive methods, based on several manifestations of the relevant capacity, all that can be established is a *ceteris paribus* law. Based on cases of uses of aspirin, all that it can be established is that *ceteris paribus*, aspirin relieves headaches. So, it is questionable that talk about capacities has extra content over talk about ordinary causal laws.

### 6.4.2.3 *Capacities and interactions*

To be fair to Cartwright, she has offered other reasons for her commitment to capacities. One of them is that capacities can explain causal interaction. She says:

> causal interactions are interactions of causal capacities, and they cannot be picked out unless capacities themselves are recognised. (Cartwright 1989, 164)

There are cases that fit this model. A venomous snake bites me and I take an antidote. The venom in my bloodstream has the capacity to kill me but I don't die because the antidote has the capacity to neutralise the venom. That's a case of causal interaction, where one capacity blocks another. I am not sure this commits us to *sui generis* capacities, as opposed to whatever chemical properties the venom and the antidote have and a law which connects these properties. But let's not worry about this. There is a more pressing problem.

Suppose that I take an aspirin while I am still hearing the continuous and desperate screaming of my daughter who suffers from colic. The aspirin has the capacity to relieve my headache, but the headache does not go away. It persists undiminished. How shall I explain this? Shall I say that this is because the screaming of my daughter has the capacity to cause aspirin-resistant headaches? This would be overly *ad hoc*. Shall I say that this is because the screaming of my daughter has the capacity to neutralise the capacity of aspirin to relieve headache? This would be very mysterious. Something has indeed happened: there has been an interaction of some sort which made aspirin not work. But why should I attribute this to a *capacity* of the screaming? If I did that, I would have to attribute to the *screaming* a number of capacities: the capacity to-let-aspirin-work-if-it-is-mild, the capacity to let-aspirin-work-if-it-is-*not*-mild-but-I-go-away-and-let-my-wife-deal-with-my-daughter, the capacity to block-aspirin's-work-if-it-is-extreme and so on. This is not exactly an argument against the role of capacities in causal interaction (though it might show that there can be causal interaction without reference to capacities). Still, it raises a genuine worry: when trying to account for causal interaction, where do we stop positing capacities and what kinds of them should we posit?

Cartwright challenges the sceptic about capacities with the following: 'the attempt to "modalise away" the capacities requires some independent characterisation of interactions; and there is no general noncircular account available to do the job' (Cartwright 1989, 164). If we could not characterise interactions without reference to capacities, we had better accept them. But why not one follow, for instance, Wesley Salmon (1997) or Phil Dowe (2000) in their thought that interactions are explained in terms of exchanges of conserved quantities? There is no compelling reason to take *them* to be capacities. We could. But then again we couldn't.

## 6.5 What are capacities?

Suppose that we do need to posit capacities. What exactly is the thing we need to posit? Cartwright is certainly in need of a more detailed account of how capacities are individuated. In her work (Cartwright 1989, 141) we are told that capacities are *of* properties and not *of* individuals: 'the property of being an aspirin carries with it the capacity to cure headaches'. But aspirin is not, strictly speaking, a property. It's something that has a property. And certainly it does not *carry* its capacity to relieve headaches in the same way in which it carries its shape or colour.

It would be more accurate to say that capacities are properties of properties. That is, that they are second-order properties. This move would create some interesting problems. It would open the way for someone to argue that capacities are functional (or causal) roles. This would imply that there must be occupants of these causal roles, which are not themselves capacities. They could be the properties (perhaps many and variable) that occupy this causal role. So, the capacity to relieve pain would be a causal role filled (or realised) by different properties (e.g., the chemical structure of paracetamol or whatever else). If, however, we take capacities to be causal roles, it would be open for someone to argue, along the lines of Prior, Pargeter and Jackson (1982) that capacities are *causally impotent*. The argument is simple. Capacities are distinct from their causal bases (since they are properties *of* them). They must have a causal basis (a realiser), since they are second-order. This causal basis constitutes a *sufficient* set of properties for the causal explanation of the manifestation of the capacity (whenever it is manifested). Hence, the capacity *qua* distinct (second-order) property is causally impotent.

Cartwright wouldn't be willing to accept this conclusion. But then capacities must be *of* properties (or be carried by properties) in a different way. What exactly this way is it is not clear. She asks: 'Does this mean that there are not one but two properties, with the capacity sitting on the shoulder of the property which carries it?' And she answers: 'Surely not' (ibid., 9). But no clear picture emerges as to what this relation of '*a* carrying *b*' is. (And is this 'carrying' another capacity, as in *a* has the capacity to carry *b*?) At a different place, we are told that capacities have powers, which they can retain or lose (in causal interactions) (cf. ibid., 163). Is that then a third-order property? A property (power) of a property (capacity) of a property (aspirin)? I don't think Cartwright wants to argue this. But what does she want to argue?

In Cartwright (1999), she comes back to these issues. Here it seems that another possibility is canvassed, namely, that properties *themselves* are capacities. It's not clear whether she takes all properties to be capacities, but it seems that she takes at least *some* to be. We are given examples such as *force* and *charge*. I am not sure I have this right, but it seems to follow from expressions such as: 'Coulomb's law describes a capacity that a body has qua charged' (ibid., 53). It also seems to follow from considering concepts such as '*attraction, repulsion, resistance, pressure, stress*, and so on' as concepts referring to capacities (ibid., 66). In any case, it seems that she aligns herself with Sydney Shoemaker's view of properties as 'conglomerates of powers' (ibid., 70). Capacities then seem to come more or less for free: 'any world with the same properties as ours would *ipso facto* have capacities in it, since what a property empowers an object to do is part of what it is to be that property' (ibid., 70). So, it seems that Cartwright adopts a causal theory of properties, where properties themselves are causal powers.

## 6.6 Capacities and laws

A number of questions crop up at this point. *First*, are all powers with which a property empowers an object constitutive of this property? And if not, how are we to draw a distinction between constitutive powers and non-constitutive ones? For instance, is the causal power of aspirin to relieve headache on a par with its causal power to produce a pleasing white image? This is not a rhetorical question. For it seems that in order to distinguish these two powers in terms of their causal relevance to something being an aspirin, we need to differentiate between those powers that are causally relevant to a certain effect, for example, relieving pain, and those powers that are not. Then, we seem to run in the following circle. We need to specify what powers are causally relevant to something being *P*. For this, we need to distinguish the effects that are brought about by *P* in two sorts: those that are the products of causally relevant powers and those that are not. But in order to do this we need first to specify what it *is* for something to be *P*.[10] That is, we need to specify what powers are causally relevant to *P*'s identity and what are not. *Ergo*, we come back to where we started. (Recall that on the account presently discussed causal powers are the *only* vehicle to specify *P*'s identity.)

*Second* question: why is it the case that some causal powers go together and others do not? Why do certain powers have a certain kind of 'causal unity', as Shoemaker (1980, 125) put it? This is a crucial question because even if every property is a cluster of powers, the converse does not hold. Electrons come with the power to attract positively charged particles and the power to resist acceleration, but they don't come with the power to be circular. And the power of a knife to cut wood does not come with the power to be made of paper. This is important because, as Shoemaker himself observes (cf. ibid.), the *concurrence* of certain powers might well be the consequence of a *law*. It might well be that laws hold some capacities together. Hence, it seems that we cannot do just with capacities. We also need laws as our building blocks.

*Third* question: Should we be egalitarian about capacities? Is the capacity to resist acceleration on a par with the capacity to become grandparent? Or with the capacity to be a table-owned-by-George-Washington? This question is different from the first above. It relates to what in the literature is called the difference between genuine changes and mere Cambridge changes. The parallel here would be a difference between genuine capacities (properties) and mere-Cambridge capacities (properties). Here again, laws are in the offing. For it can be argued that genuine capacities (properties) are those who feature in laws of nature.

I offer these questions as challenges. They do point to a certain double conclusion. On the one hand, we need to be told more about what capacities are before we start thinking seriously that we should be committed to them. On the other hand, we seem to require *laws* as well as capacities, even if we accept capacities as building blocks.

In Cartwright (1999), she wants to advance further the view that capacities are metaphysically prior to laws. She says: 'It is capacities that are basic, and laws of nature obtain – to the extent that they do obtain – on account of the capacities' (ibid., 49). She offers no formal treatment of the issue how capacities relate to laws. Instead, we are given some examples.

> I say that Newton's and Coulomb's principles describe the capacities to be moved and to produce a motion that a charged particle has, in the first case the capacity it has on account of its gravitational mass and in the second, on account of its charge (1999, 65).

If laws describe what the entities governed by them can do on account of their capacities, these capacities should be individuated, and ascribed, to entities, independently of the lawlike behaviour of the latter. But, as noted above, it is not clear that this can be done. It seems that far from being independent of laws, the property of, say, *charge* is posited and individuated by reference to the lawlike behaviour of certain types of objects: some attract each other, while others repel each other in a regular fashion. The former are said to have opposite charges, while the latter have similar charge. Cartwright (1999, 54–5) says:

> The capacity is associated with a single feature—charge—which can be ascribed to a body for a variety of reasons independent of its display of the capacity described in the related law.

This may well be true. But it does not follow that the capacity is grounded in no laws at all. Cartwright disagrees. She (ibid., 72) claims that

> (c)apacity claims, about charge, say, are made true by facts about what it is in the nature of an object to do by virtue of being charged.

Then, one would expect an informative account of what it is in the *nature* of an object to do. Specifically, one would expect that the nature of an object would determine its capacities, and would delineate what this object can and cannot do. Teller directed my attention to the following passage, in which Cartwright says:

> My use of the terms *capacity* and *nature* are closely related. When we ascribe to a feature (like charge) a generic capacity (like the Coulomb capacity) by mentioning some canonical behaviour that systems with this capacity would display in ideal circumstances, then I say that that behaviour is *in the nature of* that feature. Most of my arguments about capacities could have been put in terms of natures.... (ibid., 84–5)

It seems that Cartwright thinks there is *no* significant distinction between capacity and nature. But suppose that she followed many other friends of capacities and distinguished between capacities and natures. Fisk (1970) and Harré (1970), among others, think that an appeal to an entity's nature can *explain* why this entity has certain capacities. In particular, Harré (1970) argues that (a) discovering the nature of an entity is a matter of empirical investigation; but (b) specifying (or knowing) the exact nature of an entity is not necessary for grounding the ascription of a power to it. He links natures and capacities thus:

> There is a $\varphi$ such that something has $\varphi$, and whatever had $\varphi$ in $C$, *would have* to $G$, i.e., if something like $\alpha$ did not have $\varphi$ in $C$ it would not, indeed *could* not $G$. (ibid., 101)

The nature $\varphi$ of an entity is thereby linked with its capacity to $G$. There are many problems with this proposal.[11] But I will focus on one. What is it that makes the foregoing *counterfactual* true? It's not enough to have the circumstances $C$ and the nature $\varphi$ in order to get $G$. This is not just because $G$ could be unmanifested. Even if we thought that the power to $G$ were always manifested in circumstances $C$ with a characteristic effect $e$, there would *still* be room for asking the question: what makes it the case that $\alpha$'s being $\varphi$ in $C$ makes it produce the characteristic effect $e$? We need, that is, something to relate (or connect) all these together and the answer that springs to mind is that it is a *law* that does the trick.[12] This law might well be a brute (Humean) regularity.[13]

An advocate of natures could say that when the nature $\varphi$ is present, there is no need to posit a law in order to explain why a certain object has a characteristic effect $e$ when the circumstances are $C$. Yet this move would not really be explanatory. It would amount to taking natures to be collections of powers and this hardly explains in an interesting way why a certain nature has the capacities it does: it just equates the nature of an object with a collection of its capacities.

We are still short of a compelling reason to take capacities seriously as fundamental non-Humean constituents of the world. At any rate, even if we granted capacities, we would still need laws to (i) identify them; (ii) connect them with their manifestations; (iii) explain their stability; (iv) explain why some (but not others) occur together; (v) explain why some (but not others) obstruct the manifestation of others. It seems then that both the epistemology and the metaphysics of capacities require laws. Cartwright has argued that *capacities are enough for laws*. If the argument in the later part of this chapter has been correct, the situation is more complicated: *laws and capacities are enough for laws*.

# Part II

# Structural Realism

# 7
# Is Structural Realism Possible?

Structural Realism (SR) is meant to be a substantive philosophical position concerning what there is in the world and what can be known of it. It is *realist* because it asserts the existence of a mind-independent world, and it is *structural* because what is knowable of the world is said to be its structure only. As a slogan, the thesis is that knowledge can reach only up to the structural features of the world. This chapter unravels and criticises the metaphysical presuppositions of SR. It questions its very possibility as a substantive – and viable – realist thesis.

## 7.1 The upward path

Let the 'upward path' to SR be any attempt to begin from empiricist premises and reach a sustainable realist position. Arguing against the then dominant claims that only the phenomena ('the world of percepts') can be known and that, even if they exist, their 'objective counterparts' are unknowable, Russell (1919, 61) suggested that 'the objective counterparts would form a world having the same structure as the phenomenal world, (a fact which would allow us) to infer from the phenomena the truth of all propositions that can be stated in abstract terms and are known to be true of the phenomena'. In Russell (1927, 226–7), he stressed that only the structure, that is, the totality of formal, logico-mathematical properties, of the external world can be known, while all of its first-order properties are inherently unknown. This logico-mathematical structure, he argued, can be legitimately *inferred* from the structure of the perceived phenomena (the world of percepts). Since this inference is legitimate from an empiricist perspective, the intended conclusion, namely, that the unperceived (or unobservable) world has a certain knowable structure, will be acceptable too. How is this inference possible? Russell rested on the (metaphysical) assumption that differences in percepts are brought about by relevant differences in their causes (stimuli). This is a *supervenience principle*: if two stimuli are identical, then the resulting percepts will be identical. I call this the 'Helmholtz–Weyl' principle, for it was Hermann Helmholtz who first enunciated it: 'we are justified, when different perceptions offer themselves to us, to infer that the underlying real conditions are different' (quoted by Weyl 1963, 26). Hermann Weyl endorsed it because he thought it grounded the possibility of knowing something about the 'world of things in themselves'. Yet, what is known of this world via the Helmholtz–Weyl principle, Russell (and Weyl 1963, 25–6) thought, is its structure. For if we conjoin the Helmholtz–Weyl principle with a principle of 'spatio-temporal continuity' (that the cause is spatio-temporally continuous with the effect), Russell (1927, 226–7) said that we can have 'a great deal of knowledge as to the *structure* of stimuli'. This knowledge is that 'there is a roughly one-one relation between stimulus and percepts', which 'enables us to infer certain mathematical properties of the stimulus when we know the percept, and conversely enables us to infer the percept when we know these mathematical properties of the stimulus' (ibid.) The 'intrinsic character' of the stimuli (i.e., the nature of the causes) will remain unknown. The structural isomorphism between the world of percepts and the world of stimuli isn't enough to reveal it. For Russell, this is just as well. As he also points out: '(...) nothing in physical science ever depends upon the actual qualities' (ibid., 227). Still, he insists, we can know something about the structure of the world (cf. ibid., 254).

There may be good reasons to doubt the Helmholtz–Weyl principle (e.g., that the stimuli overdetermine the percepts). But even if we granted it, the Russellian argument that we can have *inferential* knowledge of structural isomorphism between the world of percepts and the world of stimuli requires a minor miracle. The Helmholtz–Weyl principle is not strong enough on its own to generate the required isomorphism. The determination it dictates is one-way: same stimuli, same percepts. The establishment of isomorphism requires also the converse of the Helmholtz–Weyl principle – viz., same percepts, same stimuli. Precisely because Russell doesn't have the converse principle, he talks of 'roughly one-one relation'. Yet, he has failed to motivate the claim that the relation should be 1–1. (Why can't the same stimuli produce different perceptions at different times, for instance?) Besides, does it make good sense to talk of 'roughly one-one relation'? Either it is or it isn't one–one. If it is, we have structure-transference. But if it isn't, we don't.

Given the importance of the converse of the Helmholtz–Weyl principle for the Russellian argument, does it have any independent motivation? From a realist point of view, it should be at least in principle possible that the (unobservable) world has 'extra structure', that is, structure not necessarily manifested in the structure of the phenomena. If there is such 'extra structure', the required structural relation between the phenomena and the (unobservable) world should *not* be isomorphism but embedability, namely, the phenomena are isomorphic to a substructure of the world. Then the attraction of the original Russellian attempt to motivate a compromise between empiricism and realism is lost. This attraction was that one could stay within the empiricist confines, and yet also claim that the structure of the (unobservable) world can be known (*inferred*). When isomorphism gives way to the weaker (but more realistic) requirement of embedability, there is no (deductive) constraint any more on what this extra structure of the (unobservable) world might be: the structure of the phenomena no longer dictates it. Hence, from an empiricist-cum-structuralist perspective it's a live option to invest the unobservable world with whatever extra structure one pleases, provided that the phenomena are embedded in it. Van Fraassen (1980) has just taken this option while also noting that an empiricist can consistently remain agnostic about the reality of the posited structure of the unobservable world. So, Russell's attempted compromise between empiricism and realism, based on the thought that the structure of the world can be *inferred* (and hence known) from the structure of the phenomena, collapses.

Weyl, for one, did endorse the converse of the Helmholtz–Weyl principle, namely, that '(t)he objective image of the world may not admit of any diversities which cannot manifest themselves in some diversity of perceptions' (Weyl 1963, 117). So, different stimuli, different percepts. This principle Weyl takes to be 'the central thought of idealism' (ibid.), and reckons that science should concede it: '(S)cience concedes to idealism that its objective reality is not given but to be constructed . . .'. But how can this be asserted a priori? And if it is, as a means to secure a priori the knowability of the structure of the world, it's inconsistent with the realist side of SR. For a realist it shouldn't be a priori false that there is a divergence between the structure of the world and the structure of the phenomena. Weyl's 'idealist' principle simply blocks the possibility that the world has any 'extra structure' which cannot manifest itself in experience.

To be sure, Russell (1927, 255) did stress that 'indistinguishable percepts need not have exactly similar stimuli'. He allowed that the relation between the percepts and the stimuli may be one–many and not one–one. This more reasonable approach makes the required structural compromise between empiricism and realism even more tenuous. Given an one–many relation, the structure of the percepts doesn't determine the *domain* of the stimuli. But it may not determine its *structure* either. If two or more different stimuli can give rise to the same percepts, knowledge of the structure of the percepts doesn't suffice for knowledge of the structure of the stimuli, assuming, as is natural to do, that these stimuli don't just differ in their 'intrinsic nature'. They may well differ in other structural characteristics which nonetheless don't surface in any way in the world of phenomena. So, the structural differences of stimuli cannot be inferred from the non-existent corresponding structural differences in the percepts. Hence, the more reasonable one–many approach leaves no room for inferring the structure of the unobservable world from the structure of the phenomena. For all we know, the unobservable world may differ from the world of phenomena not just in its 'intrinsic nature', but in its structure too.

The Russellian 'upward path' to SR faces an important dilemma. *Without* the converse of the Helmholtz–Weyl principle, it cannot establish the required isomorphism between the structure of the phenomena and the structure of the (unobservable) world. Hence it cannot establish the possibility of inferential knowledge of the latter. *With* the converse of the Helmholtz–Weyl principle, it guarantees knowledge of the structure of the world, but at the price of conceding – a priori – too much to idealism.

Russell's thesis was revamped by Maxwell – with a twist. Maxwell, who coined the term 'structural realism', took the Ramsey-sentence approach to exemplify the proper structuralist commitments. He advocated SR as a form of representative realism, where all first-order properties are unknowable and only higher-order properties of things are knowable. As he said: 'our knowledge of the theoretical is limited to its purely structural characteristics and (...) we are ignorant concerning its intrinsic nature' (Maxwell 1970a, 188). The structural characteristics (or properties) were taken to be those that are not, or could not be, 'direct referents of predicates' (cf. ibid.) Maxwell didn't take the so-called structural characteristics to be purely formal. He thought that formal properties, such as transitivity, are purely structural, but added that 'not all structural properties are also purely formal (...); in fact those referred to by scientific theories rarely are' (Maxwell 1970b, 188). Yet, he left us in the dark as to what these non-formal structural properties, which are referred to by theories, are. Saying that they are 'always of a higher logical type, (viz.,) properties of properties, properties of properties of properties etc.' (cf. ibid.) isn't enough to show that they are not purely formal. On the contrary, in Maxwell's Ramsey-sentence approach to structuralism, where the theoretical properties have been replaced by (second-order) bound variables, there is simply nothing left except formal properties and observable ones. In the Ramsey-sentence approach, all non-observable properties are just formal properties: ultimately, they are just logical relations between the higher-order bound variables. So, when Maxwell (cf. ibid.) says that the Ramsey-sentence 'refers by means of description to unobservable intrinsic properties' (whose intrinsic nature is otherwise unknowable), he misses the point. There is nothing in the Ramsey-sentence itself that tells us that what is referred to by the bound variables are properties (of any sort) of unobservable entities. This last construal, though consistent with the meaning of a Ramsey-sentence, is in no way dictated by it, as Carnap himself noted (cf. Psillos 1999, 53).

Maxwell, unlike Russell, takes the phenomena to be represented by the observable part of the Ramsey-sentence of a theory. The structure of the phenomena is not, on Maxwell's view, isomorphic to the structure of the unobservable world. Instead (descriptions of) the phenomena are embedded in logico-mathematical structures, which are abstracted from theories and are said to represent the structure of the unobservable world. While this difference might be thought sufficient to guarantee that Maxwell's compromise between empiricism and realism is viable, it fails to do so for the following reasons.

First, unless we smuggle in a suspect requirement of direct acquaintance, it isn't clear why the first-order properties of unobservable entities are unknowable. They are, after all, part and parcel of their causal role. If all these entities are individuated and become known via their causal role, there is no reason to think that their first-order properties, though contributing to causal role, are unknowable. Second, the most fatal objection to the Russellian programme, namely, Newman's (1928) claim that knowledge of structure is empty since it just amounts to knowledge of the cardinality of the domain of discourse, applies to Maxwell's thesis no less. We shall see the details of this objection in Chapter 9, so here I will simply summarise it: even when the domain of the structure of the stimuli is fixed, a relational structure on this domain can *always be defined* (cardinality permitting) in such a way as to guarantee isomorphism between the structure of the percepts and the structure of the stimuli. Hence, the only information encoded in the claim of structural isomorphism is that the domain of the stimuli has a certain cardinality (cf. Demopoulos and Friedman 1985; Psillos 1999, 61–9). Maxwell's Ramsey-style approach to SR fares no better than Russell's vis-à-vis the Newman objection: without further (non-structural) constraints on the range of the second-order variables of the Ramsey-sentence, if the Ramsey-sentence is empirically adequate, then it is guaranteed – by logic alone – to be true (cf. Psillos 1999, Chapter 3). Hence, the claim that the Ramsey-sentence captures the structure of the (unobservable) world is empty, unless – to repeat – non-structural constraints are placed on the range of its variables.

To sum up, the 'upward path' to structural realism is not viable. Structural isomorphism between the phenomena and the unobservable world cannot be inferred without contentious metaphysical assumptions – which compromise the realist side of the wedding. A Ramsey-sentence approach to structural realism would either have to abandon pure structuralism or else be an empty claim.[1]

## 7.2 The downward path

Let the 'downward path' to SR be any attempt to start from realist premises and construct a *weaker* realist position. Its most prominent exponent is Worrall (1989). Take realism to incorporate two conditions: the Independence Condition (IC), namely, that there is a mind-independent world (i.e., a world which *can* be essentially different from whatever is constituted by our conceptual capacities and our capacities to theorise); and the Knowability Condition (KC), namely, that this mind-independent world is knowable. (For a more detailed discussion of these two positions, see Chapter 1 above.) Call a thesis which subscribes to both conditions *Metaphysical Realism* (MR). Note that MR doesn't imply anything specific about what aspects of the world are knowable. It may be an overly optimistic thesis that everything in the world is knowable. Being realism, SR should subscribe at least to IC. If then SR is to be different from MR, it has to differ in its approach to KC. In particular, SR has to place a *principled* restriction on what aspects of the mind-independent world can be known. The claim is that only the *structure* of the world can be known. SR has two options available. Either there is something other than structure – call it X – in the world, which however cannot be known, or there is nothing else in the world to be known. On the first disjunct, the restriction imposed by SR is epistemic. Call this view *Restrictive Structural Realism* (RSR). On the second disjunct, the restriction is ontic: there is nothing other than structure to be known, because there is nothing other than structure. Call this view

*Eliminative Structural Realism* (ESR). So it appears that the stronger version of MR which SR tries to weaken is the claim that there is more to the world than its structure, and this more – the X – can be known. RSR grants that there is this extra X, but denies its knowability, whereas ESR eliminates this extra X altogether. What could this extra X be?

Let's think in terms of a metaphor. Take MR to assert that the world forms a gigantic (interpreted) 'natural' graph, where there are entities and their properties and relations to which they stand among themselves and higher-order properties and relations. It is an essential part of the metaphor that all these entities are definite: there are objective similarities and differences among things which constitute a fact-of-the-matter as to what these entities are and how they relate to each other. What would then be the parts of the graph which SR intends either to restrict epistemically or to eliminate? Here are the options about the supposed unknowable X: the objects (individuals); the (first-order) properties; the relations; the higher-order properties of properties and relations. Each level of abstraction creates a version of SR. The claim that we can know only the structure of the graph is, therefore, ambiguous. It may mean that:

(A): We can know everything but the individuals that instantiate a definite structure. Or, (B): We can know everything except the individuals *and* their first-order properties. Or (C): We can know everything except individuals, their first-order properties *and* their relations.

Notice that as we went down the line, we relegated more and more things to the 'unknowable extra X'. Where exactly do we draw the line and what – if any – are the consequent principled restrictions on KC?

Take RSR to mean (A). Then, RSR(A) implies that a proponent of MR should endorse the following thesis: if there were two interpreted structures which were exactly alike in all respects but their domains of discourse (M and M′, respectively), there would still be a fact of the matter as to which of those is the correct structure of the world (assuming of course that they 'surface' in the same way and they don't conflict with observations.) Realists may want to take sides on this issue. I don't think they have to. The only *possibly* substantive issue that remains is to *name* the individuals of the domain. How can this be a substantive issue? Since by hypothesis, the individuals in the two domains instantiate the very same *interpreted* (natural) structure, for each individual in domain M there is another individual in domain M′ such that the two individuals share exactly the same causal role. Whether or not these two individuals will be taken to be the same will depend on how one thinks about individuation. If one thought that properties require a substratum to which they inhere, one would also be open to the thought that there may well be two distinct individuals with all their properties in common. If so, the whole issue between MR and RSR(A) hinges upon the metaphysics of properties.

I don't intend to resolve this issue here.[2] Suffice it to note that RSR(A) is weaker than MR in a principled way only if MR is taken to accept (the questionable thesis) that two individuals can share all their properties and yet be different. It may be thought here that this might be a case of multiple realisation. But in this context, this case is inapplicable. For we are not talking about two systems which can be described differently on physical terms but instantiate the same higher-level structure. Rather, we are talking about one interpreted (natural) structure which is instantiated by two domains whose only difference is that they use different names for their individuals.

Suppose RSR means (B). Then RSR(B) implies that MR should endorse the following thesis: if there were two semi-interpreted structures which were exactly alike in all respects except their domains of discourse *and* the first-order properties attributed to individuals, there would still be a fact of the matter as to which of those is the correct structure of the world. This is a view that some realists would accept. But has RSR(B) come up with a principled epistemic restriction? (B) amounts to Carnapian 'relation descriptions' (1928, 20). Relation descriptions offer less information than 'property descriptions' (associated with RSR(A)). They describe an object as that which stands in certain relations to other objects, for example, 'a is the father of b', without further specifying its properties. What this object is is not thereby fully specified. Only what it is in relation to other objects is being given. Although 'relation descriptions' don't entail unique 'property descriptions', they do offer some information about an object, because, generally, they entail *some* of its properties. For instance, from the relation description 'a is the father of b' we can conclude that a is male, that a is a parent, and so on. More interestingly, from relational descriptions about, for example, electrons, we can legitimately infer the existence of some first-order properties, namely, negative charge or mass. There is simply no natural epistemic cut between the relational and the first-order properties. Hence, given a rich enough relation description, one may infer enough properties of the object to identify it to a great extent. Hence, (B) doesn't imply inherent unknowability. That some properties may remain unspecified doesn't justify the claim that only relations are knowable. Although

RSR(B) is weaker than a realist position which says that *all* properties and relations can be known, the difference is only one of degree. RSR(B) doesn't preclude any *specific* property from being knowable. Rather, it says that if we go for relation descriptions, some properties may be unknowable. MR should be happy with this.

Having shown that structural restrictions of types (A) and (B) above don't pose any principled restriction on MR, we should consider restriction (C). So, take RSR(C) to mean that only the higher-order properties of properties and relations are knowable. Here we have reached, as it were, pure structure. We can therefore talk of Carnapian 'purely structural definite descriptions' (1928, 25): not only are we ignorant of the objects and their properties, but also we don't know what their relations are. All we can know is the totality of the formal properties and relations of the structure, where these are those that can be formulated without reference to their meanings. Should the domain of a relation be finite, the structure of this relation is given by its logico-mathematical graph. This is a *complete structure description*, whose full verbal equivalent is a list of ordered tuples. RSR(C) is the only characterisation of SR which can impose a principled limitation on what is knowable. To return to the original graph metaphor, the claim is that all that can be known is the structure of the graph. Note that RSR(C) doesn't deny that there are definite relations which structure the world. It just claims that we cannot know *what* they are.

Is RSR(C) a defensible position? Even though we followed the realist 'downward path', we have just reached the highest point of the empiricist 'upward path'. All the problems met there apply here too. In particular, there are two problems. The problem of *motivation* and the problem of *content*. Why should we consider accepting the restriction RSR(C) places on what can be known of the world? And, what is the formal structure a structure *of*? If we could just infer the formal structure of the world from the phenomena, the motivational question would be answered. But RSR(C) fares no better on this score than Russellian empiricism and Maxwellian structural realism. For the reasons expressed in the previous section, there is no inferential path from the structure of the phenomena to the structure of the unobservable world. What about the problem of content? Two issues are relevant here. First, in empirical science at least we should seek more than formal structure. Knowing that the world has a certain formal structure (as opposed to a natural structure) allows no explanation and no prediction of the phenomena. Second, the claim that the world has a certain formal structure is, from a realist point of view, unexciting. *That* the world has a formal structure follows trivially from set-theory, if we take the world to be a set-theoretic entity. In fact (this is Newman's objection), it follows that the world has any formal structure we please (consistent with its cardinality). *That* it has the formal structure that corresponds to a definite natural structure (i.e., that the formal structure has a certain natural structure as its *content*) is a much more exciting claim, which however cannot be established by purely structural means. The 'downward path' to SR should at least retain the realist outlook. This cannot be retained unless there is a prior commitment to the not-purely-structural thesis that the world has a definite natural structure. That this thesis isn't purely structural follows from the fact that its defence (and possibility) requires a commitment to objective (and knowable, at least in principle) similarities and differences between natural kinds in the world.

To sum up: the 'downward path' to RSR either fails to create a sustainable restriction on MR or, insofar as it focuses on the knowability of purely formal structure, it fails to be realist enough. It can be realist by talking about 'natural structures', but then again it gives up on pure structuralism.

What about ESR? Remember that ESR is eliminative: only structure can be known, simply because there is nothing else to know. A proposal in this direction has been recently defended by James Ladyman (1998) and Steven French (1999). Ladyman urges us to take SR as a metaphysical thesis. He notes: 'This means taking structure to be primitive and ontologically subsistent' (Ladyman 1998, 420). And he adds that the empirical success of theoretical structures shouldn't be taken to 'supervene on the successful reference of theoretical terms to individual entities, or the truth of sentences involving them' (Ladyman 1998, 422). French (1999, 203) stresses that on their view 'there are no unknowable *objects* lurking in the shadows (...)'. The details of this view are not specified yet, so it is difficult to evaluate it properly. (See, however, the next chapter.)

In any case, I fail to see how the eliminativism it suggests is possible. If structures are independent of an ontology of individuals and properties, we cannot even speak of any structural relation (be it isomorphism, or embedding or what have you) between structures. I doubt that we have any special insight into structural relations – we establish them by pairing off individuals and mapping properties and relations onto one another. Even in group theory – Ladyman and French's paradigm case for their thesis – the group-structure is detached from any *particular* domain of individuals, but not from the very notion of a domain of individuals. To hypostatise structures is one thing (perhaps legitimate). But to say that they don't supervene on their elements is quite another.

It implies the wrong ontological thesis that structures require no individuals in order to exist and the wrong epistemic thesis that they can be known independently of (some, but not any in particular, set of) individuals which instantiate them. Note that if the structures 'carry the ontological weight' (ibid., 204), we can only take the identity of structures as something ontologically primitive (since the notion of isomorphism requires different domains of individuals which are paired-off). But I am not sure whether we can even make sense of this primitive structural identity. And if we introduce individuals as 'heuristic' devices (ibid.) whose sole role is 'the introduction of structures' (only to be 'kicked away' later on), we need to justify why they are just 'heuristic devices' if the *only* road to structure is via them. I conclude that it's hard to see how the ontological revision ESR suggests is possible. Besides, if 'theories tell us not about the *objects* and *properties* of which the world is made, but directly about *structure* and *relations*' (Ladyman, 1998, 422), it is a major miracle that they can be used to represent the world we live in. Unless we buy into some problematic metaphysical thesis which somehow 'constructs' the individuals out of relations, the world we live in (and science cares about) is made of individuals, properties and their relations.

Let me end with a positive note. One way to read SR is as a modest epistemic thesis which emerges from looking into the history of scientific change. There is no heavy metaphysical machinery behind it, nor absolute claims about what can or cannot be known. It is just a sober report of the fact that there has been a lot of structural continuity in theory-change: we have learned a lot about theoretical and empirical laws, although our views about what entities in the world are related thus have seen some major discontinuities. In a certain sense, this is the insight behind Worrall's motivation for SR. All this can reasonably be accepted without abandoning realism. What isn't acceptable is any form of strong thesis which draws a principled division between the (knowable) structure of the world and some forever elusive (or, worse, non-existent) X.

# 8

# *The* Structure, the *Whole* Structure and Nothing *but* the Structure?

> 'All right', said the Cat; and this time it vanished quite slowly, beginning with the end of the tail, and ending with the grin, which remained some time after the rest of it had gone.
>
> 'Well, I have often seen a cat without a grin', thought Alice; 'but a grin without a cat! It's the most curious thing I ever saw in all my life!'
>
> *Lewis Carroll, Alice's Adventures in Wonderland*

Structuralism in the philosophy of science comes in many varieties. It ranges from a methodological thesis (concerning the nature of scientific theories and claiming that they are best understood as families of models) to an ontic position (concerning what there is and claiming that structure is all there is). In between, there is an epistemic view: there is more to the world than structure, but of this more nothing but its structure can be known. In this chapter, I shall discuss the radical ontic position. As noted already towards the end of the last chapter, ontic structuralism (henceforth OS) is still quite an amorphous, though suggestive, position. The slogan is: 'all that there *is*, is structure' (da Costa and French 2003, 189). But then there are different claims of varying strengths. Here are some of them.

- Objects should be reconceptualised in 'purely structural terms' (French & Ladyman 2003a, 37).
- '(T)here are no unknowable *objects* lurking in the shadows (...)' French (1999, 203).
- Objects play only 'a heuristic role allowing for the introduction of the structures which then carry the ontological weight' (op. cit., 204).

- '(T)he only non-structural understanding of the nature of such objects is metaphysical and unwarranted by the physics itself (...)' (French & Ladyman 2003a, 45).
- '(T)here are mind-independent modal relations between phenomena (both possible and actual), but these relations are not supervenient on the properties of unobservable objects and the external relations between them, rather this structure is ontologically basic' (op. cit., 46).

There are different ways to read OS. Here are four interpretative candidates, concerning objects. Eliminative OS: there are *no* objects. Reconstructive OS: there *are* objects but are reconceptualised in a structuralist way. Formal OS: structurally reconceptualised, 'objects' are mathematical entities. Semi-formal OS: it is only unobservable 'objects' that have to be reconceptualised structurally as mathematical entities. And then there is the issue of how to understand properties and relations. Mild OS: structure is ontologically basic, being not supervenient on the intrinsic properties of objects. Radical OS: structure is ontologically basic because there are *no* objects.

Presently, I won't examine these interpretative issues. I will focus on the slogan: all that there *is*, is structure. The slogan captures the spirit of OS, namely, pure structuralism. According to the slogan, 'objects' are, at best, positions in structures. What makes French and Ladyman's OS distinctive is that it means to be a realist position: ontic structural realism. The slogan aims to refer to the mind-independent structure of the world. This structure is meant to be modal (or causal): hence, modal ontic structural realism.

## 8.1 *The* structure

Structuralists often talk about *the* structure of a certain domain. Is this talk meaningful? If we consider a domain as a set of objects, it can have any structure whatever. In particular, it can have a structure W isomorphic to another, independently given, structure W′, provided that the domain has enough objects. This can be seen in various ways. Given enough building blocks and rods, they can be so arranged that they have the structure of the London Underground or of the Paris Metro. Given that a whole can be divided into any number of parts, isomorphic structures can be defined on any two distinct wholes, for instance, a brick wall and the top of my desk. The operative notion here is the standard definition of similarity of structure: two classes A and B are similar in structure (isomorphic) iff there is an one–one correspondence *f* between the members of A and B and whenever any n-tuple <$a_1 \ldots a_n$> of members of A stand to relation P their image <$f(a_1) \ldots f(a_n)$> in B stands to relation f(P). It's a consequence of this definition that any two similar classes (i.e., any two classes with the same cardinality) can have the same structure. The upshot is that if we start with the claim that a certain domain D has an arbitrary structure W, and if we posit another domain D′ with the same cardinality as D, it follows as a matter of logic that *there is* a structure W′ imposed on D′ which is isomorphic to W. As will be explained in detail in the next chapter, this claim has been the motivating thought behind Newman's critique of Russell's structuralism and of Putnam's model-theoretic argument against metaphysical realism.

Things can be worse. Take Newtonian mechanics, where $\mathbf{F} = m\mathbf{a}$, and compare it with a reformulation of it, according to which **F** always is the vector sum of two more basic forces $\mathbf{F}_1$ and $\mathbf{F}_2$. Here we have two non-isomorphic structures, which are nonetheless, empirically equivalent. Which of them is *the* structure of the Newtonian world? Or consider the set S = {1, 2,...,12) and take R to be such that xRy if x evenly divides y. This structures the domain in a certain way: R is reflexive, anti-symmetric and transitive. But then, define R′ on S as follows: xR′y if 3 evenly divides (x–y). The structure of S is now different, since R′ is reflexive, symmetric and transitive.

*Ergo, the* structure of a domain is a relative notion. It depends on, and varies with, the properties and relations that characterise the domain. A domain has no inherent structure, unless some properties and relations are imposed on it. Or, two classes A and B may be structured by relations R and R′ respectively, in such a way that they are isomorphic, but they may be structured by relations Q and Q′ in such a way that they are *not* isomorphic.

Following the terminology introduced by Dummett (1991, 295) and Stuart Shapiro (1997, 85), let's call a 'system' a collection of objects with certain properties and relations. We may call it a 'relational system' to emphasise the fact that it's so structured that it satisfies a certain condition, for example, Peano's axioms or Newton's laws. A 'structure' then is the *abstract form* of this system. Focusing on structure allows us to abstract away all features of the objects of the system that do not affect the way they relate to one another. It is clear that the *system* comes already structured. We can then talk about its abstract *structure*, but this talk is parasitic on the system being a particular and already structured complex.

This ushers in the *basic structuralist postulate*. Among the many structures that can characterise a system some is privileged. This is *the* structure of this system as specified by the relationships among the objects of the system. It's this postulate that renders talk about *the* structure of system meaningful.

This postulate rests on a non-structural assumption. That a system has a definite structure (*this* rather than *that*) follows from the fact that certain relations and not others characterise it. But that certain relations (and not others) characterise a system is a basic non-structural feature of this system. The issue is what exactly makes a structure privileged.

Following Shapiro, we might distinguish between two versions of structuralism. *Ante rem* structuralism has it that structures are abstract, freestanding, entities: they exist independently of systems, if any, that exemplify them. In a sense, they are like universals (more like Platonic universals than Aristotelian ones). *In re* structuralism takes systems as being ontically prior to structures: it denies that structures are freestanding entities. Structures are abstractions out of particular systems and claims about *the* structure are, in effect, to be understood in one of the following ways. Talk, say, about *the* natural-number structure is talk about *any* system structured in a certain way, namely, having an infinite domain, a distinguished element $e$ in it and a successor function $s$ on it such that the conditions specified by the Peano axioms are satisfied. Or, talk about the natural-number structure is talk about *all* systems structured in the above way (cf. Reck & Price 2000). Both ways take talk about structure to be relative to systems, but the first takes *the* structure to be that of a certain system (admitting, however, that any other isomorphic system would do equally well), while the second takes *the* structure to be a generalisation over all isomorphic systems. Both ways take it that were there no systems, there would be no structures.

One important difference between *ante rem* and *in re* structuralism concerns the role of objects in structures. Since *in re* structuralism focuses on relational systems, it takes the objects of a structure to be whatever objects systems with this structure have. According to *in re* structuralism, there are no extra objects which 'fill' the structure. It's then obvious that the objects that 'fill' the *in re* structures have more properties than those determined by their interrelationships in the structure. They are given, and acquire their identity, independently of the abstract structure they might be taken to exemplify.

By hypostatising structures, *ante rem* structuralism introduces more objects: those that 'fill' the abstract structure. Of these 'new' objects nothing is asserted than the properties they have in virtue of being places (or roles) in a structure. These places cannot be identified with the objects of any or all of the *in re* structures that are isomorphic to the abstract pattern. This is what Shapiro calls 'places-are-objects' perspective. The 'fillers' of the abstract (*ante rem*) structure are places, or positions, in the structure, yet if one considers the structure in and of itself, they are genuine objects. After all, they *must* be such since the abstract structure instantiates itself (cf. Shapiro 1997, 89). Given that an instantiated abstract structure needs objects to be instantiated into, the places of the abstract structure must be objects. Mathematical structuralism, then, does not view structures without objects. It's not revisionary of the underlying ontology of objects with properties and relations.

Intermediate moral: structures need objects. This holds for both *ante rem* and *in re* structuralism. These two kinds of structuralism might need different objects, but they both need them.

The distinction between *ante rem* and *in re* structuralism may cast some light on the question noted above, namely, what makes a certain structure privileged? Let's assume, for the sake of the argument, that there are *ante rem* (freestanding) structures. Perhaps, one may argue, what makes a structure W of a system S privileged (i.e., what makes it *the* structure W of system S) is that W is isomorphic to an *ante rem* structure W′. But this thought leads to regress. For the same question can be asked about *ante rem* structure W′: what makes *this* structure privileged? If the answer is that it is isomorphic to another *ante rem* structure W″, we are led to regress. If the answer is different, some independent reason has to be given as to why an *ante rem* structure W′ is privileged. Perhaps, in pure maths the point is innocuous. Mathematicians define and study all sorts of structures and *any* structure, defined implicitly by a set of axioms, will do.

Things are more complicated when it comes to physical systems. Here *ante rem* structuralism is ill-motivated. Finding the structure of a natural system is an a posteriori (empirical) enterprise. Its structure is *in re*. And it is a natural structure in the sense that it captures the natural (causal-nomological) relations among the objects of the system. It is *the* structure that delimits a certain domain as possessing causal unity. Hence, it is grounded on the causal relations among the elements of the domain. It's these facts (that the structure is *in re* and that it confers causal unity) that make some structure privileged vis-à-vis all other structures that *can* be defined on the elements of a system. But then, it's odd to argue that a certain structure W is privileged because it is isomorphic to an abstract structure W′. First, it may not be. Given that the discovery of the *in re* structure is an empirical matter, it may not be isomorphic to any of a set of antecedently given *ante rem* structures. Second, even if the discovered *in re* structure turns out to be isomorphic to an *ante rem* one, the order of ontic priority has been reversed: the *ante rem* structure is parasitic on the *in re*; it's an abstraction from it.

Take, then, an *in re* structure W and an *ante rem* one W′ which are isomorphic. Let us add that W is the structure of a concrete physical system and that W has a certain causal unity and role. W′ instantiates itself, but since it is *ante rem*, W′ has no causal unity and plays no causal role. Yet, W and W′ are isomorphic. It follows that the causal unity and the causal role of W are not determined by its structural properties, that is the properties it shares with W′ and in virtue of which it is isomorphic to W′. If it were so determined, W′ would have exactly the same causal unity and causal role as W. But it has not. So, if we take structures to be causal, we should *not* look to their structural properties for an underpinning of this causal unity and activity. This is not surprising. Places in structures and formal relations do not cause anything at all. It's the 'fillers' of the places and concrete (*in re*) relations that do.

Consider what Ladyman (2001, 74) says: '(T)here is still a distinction between structure and non-structure: the phenomena have structure but they are not structure'. And French & Ladyman (2003b, 75) claim: 'What makes a structure "physical"? Well, crudely, that it can be related – via partial isomorphisms in our framework – to the "physical" phenomena. This is how "physical" content enters'. Claims such as these still waver between an *ante rem* and an *in re* understanding of structures. But they seem to concede the point that structuralism cannot be pure: the phenomena are able to give 'content' to a structure precisely because they are *not* themselves structure.

Moral: To be able to talk meaningfully about *the* structure (of anything) OS needs to respect the basic structuralist postulate. This compromises pure structuralism. OS will take structures to be either *ante rem* or *in re*. Objects are needed in either case. If OS reifies *ante rem* structures, their causal unity, role and efficacy are cast to the wind. If OS gives ontic priority to *in re* structures, there is more to the world than structure.[1]

## 8.2 The *whole* structure

Can the structure of a domain be known and is it necessary for science to discover the whole of it? Structural empiricism (one form of which is constructive empiricism) allows that the structure of appearances can be known, but denies that science should or need aim at knowing more. Structural empiricism stresses that a theory is successful if the structure of appearances is isomorphically embedded in a model of the theory. This allows that the theory is empirically adequate if it captures *just* the (abstract) structure of appearances. But there is nothing in structural empiricism that prohibits it from arguing that the *in re* structure of appearances is *identical* to the empirical sub-model of the theory. Structural empiricism (and in particular constructive empiricism) is not pure structuralism. It's not revisionary of the idea that appearances consist of (observable) objects with (observable) properties and relations: it takes appearances to be *in re* structures. These *in re* structures are knowable (at least in principle). If they were not, no theory could save them.

Structural empiricism takes scientific theories literally and rests on the notion of truth as correspondence. This means that it takes seriously the possibility that the world might well have excess structure over the appearances. After all, the world *must* have excess structure if one of the many incompatible theoretical models of the appearances is to be correct. But for this excess structure to exist it should be causally connected to the structure of appearances. Perhaps, this excess structure might exist in complete causal isolation. But this thought would be revisionary of actual science. And it would leave totally unmotivated the view that theories should be taken at face value, as telling a causally unified story as to how the world might be. A scientific theory does not describe two causally disconnected systems (the system of appearances and the system of what happens 'behind' them). Rather, it tells a story as to how these two systems are causally connected. Structural empiricism should at least leave it open that the excess structure of the world is causally connected to the structure of appearances. The price for this is that structural empiricism buys into a substantive metaphysical assumption: that it's at least possible that the structure of appearances is *causally connected* to a deeper unobservable structure.

(Epistemic) structural realism of the form discussed in the previous chapter is more optimistic than structural empiricism. It claims that the structure of the world behind the appearances is s knowable. As already noted in Chapter 7, Section 7.1, structural realism, in its Russellian stripe, claims that there is an inferential route, from the structure of appearances to the structure of their (unobservable) causes, based on the claim that the appearances and their causes have the same structure. But, as noted in the previous chapter, this programme fails on many counts. In its Maxwellian stripe, structural realism improves on the Russellian version by denying the inferential route: the world has excess structure over the appearances, but this excess structure can be captured (hypothetico-deductively, as it were) by the Ramsey-sentence of an empirically adequate theory.

The chief problem with this view is that, on a Ramsey-sentence account of theories, it turns out that an empirically adequate theory *is* true (cf. Psillos 1999, 61–9 & the next chapter, Sections 9.4 & 9.5). The supposed 'excess' structure of the world turns out to be illusory. One point brought home by the discussion over constructive empiricism is that we should take seriously the idea that the world may have 'excess structure' over the appearances. A theory describes a way the world might be and an empirically adequate theory might be false: the world might be different from the way it is described by an empirically adequate theory. This idea is not honoured by Ramsified structural realism, unless it drives a wedge between empirical adequacy and truth. This wedge can be driven in position only if it is accepted that the world has already built into it a natural structure, a structure that the Ramsey-sentence of the theory might *fail* to capture. This is a non-structural principle. And since we are talking about the world, this has to be an *in re* structure.

Structural realism in both of its foregoing stripes is not pure structuralism. It treats the world as a relational system with objects and properties and relations. Its point is epistemic rather than ontic. But in the end, the epistemic claim that *only* structure can be known comes to nothing. To cut a long story short, given that we talk about *in re* structures, there are objects that 'fill' the structures, these objects have properties over and above those that are determined by their interrelationships within the structure (at least) some of these (non-structural) properties are knowable (e.g., that they are not abstract entities, that they are in space and time, that they have causal powers etc.) and, in any case, these *in re* structures are individuated by their non-structural properties since it's in virtue of these (non-structural) properties that they have causal unity and are distinguished from other *in re* structures.

Moral: Epistemic structural realism promises that the 'excess structure' of the world can be known, but fails to deliver on its promise unless more than structure can be known.

## 8.3 And *nothing but* the structure

Ontic structuralism is meant to be a substantive thesis. The structure of the world (which, presumably, is all there is) is a *causal* structure. French and Ladyman (2003b, 75) write: '(...) causal relations constitute a fundamental feature of the structure of the world'. But OS cannot accommodate causation within the structuralist slogan.

There is a *prima facie* promising way to understand causation in a structuralist framework: we can think of it as structure-persistence or structure-preservation. This approach has been captured in Russell's (1948) structural postulate: events (complex structures) form causal chains, where the members of the chain are similar in structure. The idea is that causation involves structural persistence without qualitative persistence. However, as Russell recognised, there are causal changes that do not involve structure-persistence, for example, the explosion of a bomb. Besides, we need to specify more precisely what exactly it is that persists. For in any process, and with enough ingenuity, *something* that persists can always be found. We need, therefore, an account of those characteristics of a process whose persistence renders this process causal. The natural candidate for such an account should involve objects and their properties.

To see that structural persistence is not enough for causation consider what I call the 'which is which' problem. Suppose that a causal chain consists in some kind of structural similarity between events *c* and *e*. Suppose also that we accept the strong view that this structural similarity is all there is to causation. Which then is the cause and which the effect? Structural considerations alone cannot afford us a distinction: cause and effect are isomorphic. A corollary of this is that structural considerations alone cannot distinguish between a case of persistence (where an event persists over time) and a case of change (where an event causes another event to happen). Both cases are structurally identical if causation consists in structural continuity. Another corollary is that structural considerations cannot distinguish between two isomorphic (but qualitative distinct) systems which of them is the cause of a certain event. One might try to avoid these by taking structures to be in space and time and by arguing that the cause is the structure that precedes in time the other structures. This would take structures to be *in re*. Only concrete systems can be in space and time. Being in space and time are *not* structural properties. So non-structural properties are necessary for causal relations.

But even if we leave all this behind, we can still question the *rationale* for taking causation to be exhausted by a chain of isomorphic structures. If causation is a relation of dependence between events (where dependence can be understood in any of the standard ways: nomological, counterfactual and probabilistic),[2] then it should be clear that the idea of isomorphism between cause and effect is undermined. There is nothing in causation-as-dependence which dictates that cause and effect should share structure. If '*c* causes *e*' is understood in any of the above dependence-senses (e.g., *e* counterfactually depends on *c* and the like), *c* and *e* may have any structure whatever. What if we take causation as a productive relation between events? If we take the cause to produce the effect, or if we think there is a mechanism that connects cause and effect, we might also think that structure-persistence or structure-transference offers the local tie that links cause and effect.

Note an irony. If we take this line, that *c* causes *e* depends on a non-structural principle. The relation of transference of structure from one event to another is not structural. Two events (or systems) may have the same structure though they are not causally connected. That this structural similarity is due to a causal connection between them is a non-structural claim. It cannot depend solely on the structural properties of the events (or systems). Though critical of the structuralist metaphysics, Chakravartty (2003, 873) suggests that OS might take causation to be brute regularity: one structure follows the other. Yet, OS cannot have it both ways. If it goes for Chakravartty's suggestion, it can no longer claim that causation consists in structure-preservation or transference. Besides, if it goes for Chakravartty's suggestion, it will inherit all the problems with understanding modality within a regularity account of causation.

Another way to highlight the problems that OS has with causation concerns the causal relata. Standardly, the relata are taken to be either events or facts. On either view, causal relations depend on objects having properties and standing in relations to each other. Perhaps, a Davidsonian view of events might seem congenial to OS: events are *particulars* that can be described in a number of ways. But Davidsonian events are *in re*: they are in space and time. Hence, they cannot be abstract structures. Notice, *a propos*, that there is an interesting but innocuous way to understand the structuralist claim. Most objects (with the exception of fundamental particles) are structured complexes. Events or facts involving these objects will involve their structure. Consequently, their structure (e.g, the structure of a molecule) is causally relevant to what these objects can do. But these are *in re* structures: they depend on the properties and relations these objects have. There is no entry point for OS here. To put the point bluntly: the truth-makers of causal claims require objects and properties.

Could ontic structuralism adopt *causal structuralism*? As John Hawthorne (2001) explains it, causal structuralism is the view that all there is to properties is their causal profile, that is the causal powers they confer on their possessors. It is a structuralist view because it denies quidditism, the view that there is something to a property – a *quiddity* – over and above its causal profile. We may think of causal structuralism as the view that properties have no intrinsic nature over and above their causal profile. So, for every (non-logical or non-mathematical) property, there *isn't* its causal role (profile) and its 'role filler'; there is only its causal role. If OS is taken to be causal structuralism, it amounts to the denial of quidditism. Is this, however, progress? First, it's not obvious that quidditism is wrong.[3] But suppose it is. Causal structuralism does not eliminate or avoid properties. Though, it dispenses with their quiddities, it accommodates properties and secures their existence and causal efficacy via their causal profile. OS would in fact require a kind of causal *hyperstructuralism*, whereby causal profiles are purely structural. But then we end up with nothing but formal structure, with no substantive properties and relations to tell us what this structure is, how it causes anything to happen etc. Second, causal structuralism would commit OS to a substantive account of causation, where causal facts are determined by the causal powers of properties. But this account of causation cannot be purely structural or formal. Causal facts would depend on the causal powers themselves and not on their structure or formal properties. The bottom line, I think, is that causal structuralism is at odds with the slogan that structure is all there is.[4]

Moral: By going modal, ontic structuralism promises to close the gap between abstract *ante rem* structures and concrete *in re* ones. But the modal features of the world are not purely structural. Nor can causation be anything like 'the cement of the universe' if structure is all there is. Worse, we cannot make sense of causation if structure is all there is.

# 9
# Ramsey's *Ramsey-sentences*

Ramsey's posthumously published *Theories* has become one of the classics of the twentieth century philosophy of science. The paper was written in 1929 and was first published in 1931, in a collection of Ramsey's papers edited by Richard Braithwaite. *Theories* was mostly ignored until the 1950s, though Ramsey's reputation was growing fast, especially in relation to his work in the foundations of mathematics. Braithwaite made *some* use of it in his *Scientific Explanation*, which appeared in 1953. It was Hempel's *The Theoretician's Dilemma*, published in 1958, which paid Ramsey's paper its philosophical dues. Hempel coined the now famous expression 'Ramsey-sentence'.

When Rudolf Carnap read a draft of Hempel's piece in 1956, he realised that he had recently *re-invented* Ramsey-sentences. Indeed, Carnap had developed an 'existentialised' form of scientific theory. In the protocol of a conference at Los Angeles, organised by Herbert Feigl in 1955, Carnap is reported to have extended the results obtained by William Craig to 'type theory, (involving introducing theoretical terms as auxiliary constants standing for existentially generalised functional variables in "long" sentence containing only observational terms as true constants)' (Feigl Archive, 04-172-02, 14). I have told this philosophical story in some detail in Psillos (1999, Chapter 3). I have also discussed Carnap's use of Ramsey-sentences and its problems (see Psillos 1999, Chapter 3; 2000b; 2000c).

In the present chapter I do two things. First, I discuss Ramsey's own views of Ramsey-sentences. This is an important issue not just (or mainly) because of its historical interest. It has a deep philosophical significance. Addressing it will enable us to see what Ramsey's lasting contribution to the philosophy of science was as well as its relevance to today's problems. Since the 1950s, where the interest in Ramsey's views has mushroomed, there have been a number of different ways to read Ramsey's views and to reconstruct Ramsey's project. The second aim of the chapter is to discuss the most significant and controversial of this reconstruction, namely, structuralism. After some discussion of the problems of structuralism in the philosophy of science, as this was exemplified in Russell's and Maxwell's views and has re-appeared in Zahar's and Worrall's thought, I argue that, for good reasons, Ramsey did not see his Ramsey-sentences as part of some sort of structuralist programme. I close with an image of scientific theories that Ramsey might have found congenial. I will call it *Ramseyan humility*.[1]

## 9.1 Ramsey's *Theories*

*Theories* is a deep and dense paper. There is very little in it by way of stage-setting. Ramsey's views are presented in a compact way and are not contrasted with, or compared to, other views. In this section, I offer a brief presentation of the main argumentative strategy of *Theories*.[2]

Ramsey's starting point is that theories are meant to explain facts, those that can be captured within a 'primary system' (Ramsey 1931, 212). As an approximation, we can think of it as the set of all singular observational facts and laws. The 'secondary system' is the theoretical construction; that part of the theory which is meant to explain the primary system. It is a set of axioms and a 'dictionary', that is 'a series of definitions of the functions of the primary system (...) in terms of those of the secondary system' (ibid., 215). So conceived, theories entail general propositions of the primary system ('laws'), as well as singular statements ('consequences'), given suitable initial conditions. The 'totality' of these laws and consequences is what 'our theory asserts to be true' (ibid.).

This is a pretty standard hypothetico-deductive account of theories. Ramsey goes on to raise three philosophical questions. Here are the first two:

(1) Can we say anything in the language of this theory that we could not say without it? (ibid., 219)
(2) Can we reproduce the structure of our theory by means of explicit definitions within the primary system? (ibid., 220)

The answer to the first question is *negative* (cf. ibid., 219). The secondary system is eliminable in the sense that one could simply choose to stick to the primary system without devising a secondary system in the first place. The answer to the second question is *positive* (cf. ibid., 229). But Ramsey is careful to note that this business of explicit definitions is not very straightforward. They are indeed possible, but only if one does not care about the complexity or arbitrariness of these definitions. The joint message of Ramsey's answers is that theories need *not* be seen as having excess content over their primary systems.

These answers point to two different ways in which this broadly anti-realist line can be developed. The first points to an eliminative instrumentalist way, pretty much like the one associated with the implementation to theories of Craig's theorem. Theoretical expressions are eliminated *en masse* syntactically and hence the problem of their significance does not arise. The second answer points to a reductive empiricist way, pretty much like the one associated with the early work of the Logical Empiricists – before their semantic liberalisation. Theoretical expressions are not eliminated; nor do they become meaningless. Yet, they are admitted at no extra ontological cost.

So far, Ramsey has shown that the standard hypothetico-deductive view of theories is *consistent* with certain anti-realist attitudes towards their theoretical part. He then raises a third question:

(3) (Are explicit definitions) *necessary* for the legitimate use of the theory? (ibid., 229)

This is a crucial question. If the answer is positive, some form of anti-realism will be the necessary bed-fellow of the hypothetico-deductive view. But the answer is negative: 'To this the answer seems clear that it cannot be necessary, or a theory would be no use at all' (ibid., 230). Ramsey offers an important *methodological* argument against explicit definitions. A theory of meaning based on explicit definitions does not do justice to the fact that theoretical concepts in science are open-ended: they are capable of applying to new situations. In order to accommodate this feature, one should adopt a more flexible theory of meaning, in particular, a theory which is consistent with the fact that a term can be meaningfully applied to new situations *without* change of meaning (cf. ibid., 230).

The important corollary of the third answer is that hypothetico-deductivism is *consistent* with the view that theories have *excess content* over their primary systems. So the possibility of some form of *realism* is open. It is of significance that Ramsey arrived at this conclusion by a methodological argument: the legitimate use of theories makes explicit definitions unnecessary. The next issue then is what this excess content consists in. That is, what is it that one can be a realist about?[3] This, I suggest, is the problem that motivates Ramsey when he writes:

> The best way to write our theory seems to be this (∃ α, β, γ) : dictionary axioms. (ibid., 231)

Ramsey introduces this idea with a fourth question:

(4) Taking it then that explicit definitions are not necessary, how are we to explain the functioning of the theory without them?

Here is his reply:

> Clearly in such a theory judgement is involved, and the judgement in question could be given by the laws and consequences, the theory being simply a language in which they are clothed, and which we can use without working out the laws and consequences. (ibid., 231)

*Judgements* have content: they can be assessed in terms of truth and falsity. Theories express judgements and hence they can be assessed in terms of truth and falsity. Note the *could* in the above quotation. It is not there by accident, I suggest. Ramsey admits that the content of theory *could* be equated with the content of its primary system. Since the latter is truth-evaluable, it can express a judgement. But this is *not* the only way. There is also the way ('the best way') he suggests: write the theory with existential quantifiers in the front.

## 9.2 Existential judgements

Ramsey's observation is simple but critical: the excess content of the theory is seen when the theory is formulated as expressing an existential judgement. In his *Causal Qualities,* Ramsey wrote: 'I think perhaps it is true that the theory of general and existential judgements is the clue to everything' (ibid., 261). In his *Mathematical Logic,* Ramsey (1931, 67ff) spent quite some time criticising Weyl's and David Hilbert's views of existential claims. Both of them, though for different reasons, took it that existential claims do not express judgements. Being an intuitionist, Weyl took it that existential claims are meaningless *unless* we possess a method of constructing one of their instances. Hilbert, on the other hand, took them to be ideal constructions which, involving as they do the notion of an infinite logical sum, are meaningless. Ramsey subjected both views to severe criticism. Its thrust is that existential propositions can, and occasionally do, express *all* that one does, or might ever be able to, know about a situation. This, Ramsey said, is typical in mathematics, as well as in science and in ordinary life. As he says: '(...) it might be sufficient to know that there is a bull somewhere in a certain field, and there may be no further advantage in knowing that it is this bull and here in the field, instead of merely a bull somewhere' (ibid., 73).

Ramsey criticised Hilbert's programme in mathematics because he thought wrong the idea that mathematics was symbol-manipulation. He did not deny that Hilbert's programme was *partly* true, but stressed that this could not be the 'whole truth' about mathematics (ibid., 68). He was even more critical of an extension of Hilbert's programme to 'knowledge in general' (i.e., to scientific theories too) (cf. ibid., 71). As we have seen, Ramsey took theories to be meaningful existential constructions (judgements), which could be evaluated in terms of truth and falsity.

The extension of Hilbert's programme to science had been attempted by Schlick (1918/1925, 33–4). He saw theories as formal deductive systems, where the axioms implicitly define the basic concepts. He thought implicit definitions divorce the theory from reality altogether: theories 'float freely'; 'none of the concepts that occur in the theory designate anything real (...)' (1918/1925, 37). Consequently, he accepted the view that the 'construction of a strict deductive science has only the significance of a game with symbols' (ibid.) Schlick was partly wrong, of course. An implicit definition is a kind of *indefinite description*: it defines a whole class (classes) of objects which can realise the formal structure, as this is defined by a set of axioms. Schlick did not see this quite clearly. But he did encounter a problem. Theories express judgements; judgements designate facts (1918/1925, 42); a true judgement designates a set of facts *uniquely* (1918/1925, 60); but implicit definitions fail to designate anything uniquely; so if seen as a network of implicit definitions of concepts a theory fails to have any factual content. This is an intolerable consequence. Schlick thought it is avoided at the point of application of the theory to reality. This application was taken to be partly a matter of observations and partly a matter of convention (cf. 1918/1925, 71).

In Schlick (1932), he came back to this view and called it the geometrisation of physics:

> by disregarding the meaning of the symbols we can change the concepts into variables, and the result is a system of propositional functions which represent the pure structure of science, leaving out its content, separating it altogether from reality. (ibid., 330)

Seen in this structuralist light, the predicate letters and other constants that feature in the axioms should really be taken to be *genuine* variables. What matters is not the meaning of these non-logical constants, but rather the deductive – hence structural – relations among them. Scientific theories are then presented as logical structures, logical implication being the generating relation. The hypothetical part comes in when we ask how, if at all, this system relates to the world. Schlick's answer is that when one presents a theory, one makes a hypothetical claim: *if* there are entities in the world which satisfy the axioms of the theory, then the theory describes these entities (cf. ibid., 330–1).

Against the backdrop of Schlick's approach, we can now see clearly Ramsey's insight. We need not divorce the theory from its content, nor restrict it to whatever can be said within the primary system, *provided* that we treat a theory as an existential judgement. Like Schlick, Ramsey does treat the propositional functions of the secondary system as variables. But, in opposition to Schlick, he thinks that advocating an empirical theory carries with it a claim of *realisation* (and not just an if–then claim): *there are* entities which satisfy the theory. This is captured by the existential quantifiers with which the theory is prefixed. They turn the axiom-system from a set of open formulas into a set of sentences. Being a set of sentences, the resulting construction is truth-valuable. It carries the commitment that not all statements of the form '$\alpha$, $\beta$, $\gamma$ stand to the elements of the primary system in the relations specified by the dictionary and the axioms' are false. But of course, this ineliminable general commitment does not imply any *specific* commitment to the values of $\alpha$, $\beta$, $\gamma$. (This last point is not entirely accurate, I think. Because it's crucial to see in what sense it is inaccurate, I shall discuss it in some detail in Section 9.7.)

## 9.3 Ramsey-sentences

As the issue is currently conceived, in order to get the Ramsey-sentence $^{R}TC$ of a (finitely axiomatisable) theory TC we conjoin the axioms of TC in a single sentence, replace all theoretical predicates with distinct variables $u_i$, and then bind these variables by placing an equal number of existential quantifiers $\exists u_i$ in front of the resulting formula. Suppose that the theory TC is represented as $TC(t_1, \ldots, t_n; o_1, \ldots, o_m)$, where TC is a purely logical m + n–predicate. The Ramsey-sentence $^{R}TC$ of TC is: $\exists u_1 \exists u_2 \ldots \exists u_n TC(u_1, \ldots, u_n; o_1, \ldots, o_m)$. For simplicity let us say that the T-terms of TC form an n-tuple $t = \langle t_1, \ldots, t_n \rangle$, and the O-terms of TC form an m-tuple $o = <o_1, \ldots, o_m>$. Then, $^{R}TC$ takes the more convenient form: $\exists uTC(u,o)$.

I will follow customary usage and call Ramsey's existential-judgements Ramsey-sentences. This is, I think, partly misleading. I don't think Ramsey thought of these existential judgements as *replacements* of existing theories or as capturing their *proper* content (as if there were an *improper* content, which was dispensable). Be that as it may, Ramsey-sentences have a number of important properties. Here they are.

$^{R}TC$ is a logical consequence of TC.
$^{R}TC$ mirrors the deductive structure of TC.
$^{R}TC$ has exactly the same first-order observational consequences as TC. So $^{R}TC$ is empirically adequate iff TC is empirically adequate.
$TC_1$ and $TC_2$ have incompatible observational consequences iff $^{R}TC_1$ and $^{R}TC_2$ are incompatible (Rozeboom 1960, 371).
$TC_1$ and $TC_2$ may make incompatible *theoretical* assertions and yet $^{R}TC_1$ and $^{R}TC_2$ be compatible (cf. English 1973, 458).
If $^{R}TC_1$ and $^{R}TC_2$ are compatible with the same observational truths, then they are compatible with each other (cf. English 1973, 460; Demopoulos 2003a, 380).

Let me sup up Ramsey's insights. First, a theory need *not* be seen as a summary of what can be said in the primary system. Second, theories, *qua* hypothetico-deductive structures, have excess content over their primary systems, and this excess content is seen when the theory is formulated as expressing an existential judgement. Third, a theory need *not* use names in order to refer to anything (in the secondary system). Existentially bound variables can do this job perfectly well.[4] Fourth, a theory need *not* be a definite description to be (a) truth-valuable, (b) ontically committing, and (c) useful. Uniqueness of realisation (or satisfaction) is not necessary for the above. Fifth, if we take a theory as a *dynamic* entity (something that can be improved upon, refined, modified, changed, enlarged), we are better off if we see it as a *growing existential sentence*. This last point is particularly important for two reasons.

The first is this. A typical case of scientific reasoning occurs when two theories $TC_1$ and $TC_2$ are conjoined ($TC_1$ & $TC_2 = TC$) in order to account for some phenomena. If we take them to be Ramsey-sentences, then $\exists uTC_1(u, o)$ and $\exists uTC_2(u, o)$ do not entail $\exists uTC(u, o)$. Ramsey is aware of this problem. He solves it by taking scientific theories to be growing existential sentences. That is to say, the theory is already in the form ∃uTC(u, o) and all further additions to it are done *within* the scope of the original quantifiers. To illustrate the point, Ramsey uses the metaphor of a fairy tale. Theories tell stories of the form: 'Once upon a time there were entities such that …'. When these stories are modified, or when new assertions are added, they concern the original entities, and hence they take place *within* the scope of the original 'once upon a time'.

The second reason is this. Ramsey never said that the distinction between primary and secondary system was static and fixed. So there is nothing to prevent us from replacing an existentially bound variable by a name or by a constant (thereby moving it into the primary system), if we come to believe that we know what its value is. His *Causal Qualities* is, in a sense, a sequel to his *Theories*. There, Ramsey characterises the secondary system as 'fictitious' and gives the impression that the interest in it lies in its being a mere systematiser of the content of the primary system. But he ends the paper by saying this:

> Of course, causal, fictitious, or 'occult' qualities may cease to be so as science progresses. E.g., heat, the fictitious cause of certain phenomena (…) is discovered to consist in the motion of small particles. So perhaps with bacteria and Mendelian characters or genes. This means, of course, that in later theory these parametric functions are replaced by functions of the given system. (Ramsey 1931, 262)

In effect, Ramsey says that there is no principled distinction between fictitious and non-fictitious qualities. If we view the theory as a growing existential sentence, this point can be accommodated in the following way. As our knowledge of the world grows, the propositional functions that expressed 'fictitious' qualities (and were replaced by existentially bound variables) might well be taken to characterise known quantities and hence be re-introduced in the growing theory as names (or constants).[5]

Viewing theories as existential judgements solves another problem that Ramsey faced. He took causal laws (what he called 'variable hypotheticals') not to be proper propositions. As he famously stated: 'Variable hypotheticals are not judgements but rules for judging "If I meet a $\phi$, I shall regard it as a $\psi$"' (1931, 241). Yet, he also took the secondary system to comprise variable hypotheticals (cf. ibid., 260). Taking the theory as an existential judgement allows Ramsey to show how the theory as a whole can express a judgement, though the variable hypotheticals it consists of, if taken in isolation from the theory, do *not* express judgements. The corollary of this is a certain meaning wholism. The existential quantifiers render the hypothetico-deductive structure truth-valuable, but the consequence is that no 'proposition' of this structure has meaning apart from the structure.[6]

## 9.4 Russell's structuralism

Let us examine the link, if any, between Russell's structuralism and Ramsey's existential view of theories. (We have had a preliminary discussion of this issue in Chapter 7, Section 7.1.)

In *The Analysis of Matter,* Russell aimed to reconcile the abstract character of modern physics, and of the knowledge of the world that this offers, with the fact that all evidence there is for its truth comes from experience. To this end, he advanced a *structuralist* account of our knowledge of the world. According to this, only the structure, that is, the totality of formal, logico-mathematical properties, of the external world can be known, while all of its intrinsic properties are inherently unknown. This logico-mathematical structure, he argued, can be legitimately *inferred* from the structure of the perceived phenomena (the world of percepts) (cf. Russell 1927, 226–7). Indeed, what is striking about Russell's view is this claim to *inferential knowledge* of the structure of the world (of the stimuli), since the latter can be shown to be *isomorphic* to the structure of the percepts. He was quite clear on this:

> (...) whenever we infer from perceptions, it is only structure that we can validly infer; and structure is what can be expressed by mathematical logic, which includes mathematics. (ibid., 254)

Russell capitalised on the notion of structural similarity he himself had earlier introduced. Two structures $M$ and $M'$ are isomorphic iff there is an 1–1 mapping $f$ (a *bijection)* of the domain of $M$ onto the domain of $M'$ such that for any relation $R$ in $M$ there is a relation $R'$ in $M'$ such that $R(x_1 \ldots x_n)$ iff $R'(fx_1 \ldots fx_n)$. A structure ('relation-number') is then characterised by means of its isomorphism class. Two isomorphic structures have identical logical properties (cf. ibid., 251).

How is Russell's inference possible? As we have already seen, Russell relied on the causal theory of perception: physical objects are the causes of perceptions.[7] This gave him the first assumption that he needs, namely, that there are physical objects which cause perceptions. Russell used two more assumptions. One is what in Chapter 7, I called the 'Helmholtz–Weyl' principle, namely, that different percepts are caused by different physical stimuli. Hence, to each type of percept there corresponds a type of stimuli. The other assumption is a principle of 'spatio-temporal continuity' (that the cause is spatio-temporally continuous with the effect). From these Russell concluded that we can have 'a great deal of knowledge as to the *structure* of stimuli'. This knowledge is that

> there is a roughly one-one relation between stimulus and percepts, (which) enables us to infer certain mathematical properties of the stimulus when we know the percept, and conversely enables us to infer the percept when we know these mathematical properties of the stimulus. (Russell 1927, 227)

The 'intrinsic character' of the stimuli (i.e., the nature of the causes) will remain unknown. The structural isomorphism between the world of percepts and the world of stimuli isn't enough to reveal it. But, for Russell, this is just as well. For as he claims: '(...) nothing in physical science ever depends upon the actual qualities' (ibid., 227). Still, he insists, we can know something about the structure of the world (cf. ibid., 254). Here is an example he uses. Suppose that we hear a series of notes of different pitches. The structure of stimuli that causes us to hear these notes must be such that it too forms a series 'in respect of some character which corresponds causally with pitch' (ibid., 227).

The three assumptions that Russell uses are already very strong but, actually, something more is needed for the inference to go through. As already noted in Chapter 7, the establishment of isomorphism requires also the converse of the Helmholtz–Weyl principle – namely, different stimuli cause different percepts. Hence, to each type of stimuli there corresponds a type of percept. If the converse of the Helmholtz–Weyl principle is not assumed, then the isomorphism between the two structures cannot be *inferred*. For, the required 1–1 correspondence between the domains of the two structures is not shown.[8]

The notion of structural similarity is purely logical and hence we need not assume any kind of (Kantian) intuitive knowledge of it. So an empiricist can legitimately appeal to it. It is equally obvious that the assumptions necessary to establish the structural similarity between the two structures are not logical but substantive. I am not going to question these assumptions here, since I did discuss them in Chapter 7.[9] What is important is that Russell's structuralism has met with a fatal objection, due to Newman (1928): the structuralist claim is *trivial* in the sense that it follows logically from the claim that a set can have *any* structure whatever, consistent with its cardinality. So the actual content of Russell's thesis that the structure of the physical world can be known is exhausted by his positing a set of physical objects with the right cardinality. The supposed extra substantive point, namely, that of *this* set it is also known that it has structure *W*, is wholly insubstantial. The set of objects that comprise the physical world cannot possibly fail to possess structure *W* because, if seen just as a set, it possesses *all* structures which are consistent with its cardinality. Intuitively, the elements of this set can be arranged in ordered n-tuples so that set exhibits structure *W*.[10]

Newman sums this up by saying:

> Hence the doctrine that *only* structure is known involves the doctrine that *nothing* can be known that is not logically deducible from the mere fact of existence, except ('theoretically') the number of constituting objects. (ibid., 144)

Newman's argument has an obvious corollary: the redundancy of the substantive and powerful assumptions that Russell used in his argument that the structure of the world can be known inferentially. These assumptions give the impression that there is a substantive proof on offer. But this is not so.

## 9.5 Maxwell's structuralism

We have already seen in Chapter 7, Section 7.1 that Maxwell associated Russell's thesis with Ramsey-sentences. It turns out that this association of Russell's structuralism with Ramsey's views (cf. Maxwell 1970b, 182) is, at least partly, wrong. To see this, recall that Russell's structuralism attempted to provide some *inferential* knowledge of the structure of the world: the structure of the world is isomorphic to the structure of the appearances. I think it is obvious that Ramsey-sentences cannot offer this. The structure of the world, as this is depicted in a Ramsey-sentence, is not isomorphic to, nor can it be inferred from, the structure of the phenomena which the Ramsey-sentence accommodates. The distinctive feature of the Ramsey-sentence $^{R}$TC of a theory TC is that it preserves the logical *structure* of the original theory. We may say then that when one accepts $^{R}$TC, one is committed to (a) the observable consequences of the original theory TC; (b) a certain logico-mathematical structure in which (descriptions of) the observable phenomena are deduced; and (c) certain abstract existential claims to the effect that there are (non-empty classes of) entities which satisfy the (non-observational part of the) deductive structure of the theory. In this sense, we might say that the Ramsey-sentence, if true, gives us knowledge of the structure of the world: *there is a certain structure which satisfies the Ramsey-sentence and the structure of the world (or of the relevant worldly domain) is isomorphic to this structure.*[11] I suppose this is what Maxwell really wanted to stress when he brought together Russell and Ramsey.

The problem with Maxwell's move is that it falls prey to a Newman-type objection. The existential claim italicised above follows logically from the fact that the Ramsey-sentence is empirically adequate, subject to certain cardinality constraints. In other words, subject to cardinality constraints, if the Ramsey-sentence is empirically adequate, it is true. The proof of this has been given in different versions by several people.[12] Its thrust is this. Take $^{R}$TC to be the Ramsey-sentence of theory TC. Suppose $^{R}$TC is empirically adequate. Since $^{R}$TC is consistent, it has a model. Call it $M$. Take $W$ to be the 'intended' model of TC and assume that the cardinality of $M$ is equal to the cardinality of $W$. Since $^{R}$TC is empirically adequate, the observational sub-model of $M$ will be identical to the observational sub-model of $W$. That is, both the theory TC and its Ramsey-sentence $^{R}$TC will 'save the (same) phenomena'. Now, since $M$ and $W$ have the same cardinality, we can construct an 1–1 correspondence $f$ between the domains of $M$ and $W$ and *define* relations $R'$ in $W$ such that for any theoretical relation $R$ in $M$, $R(x_1 \ldots x_n)$ iff $R'(fx_1 \ldots fx_n)$. We have induced a structure-preserving mapping of $M$ on to $W$; hence, $M$ and $W$ are isomorphic and $W$ becomes a model of $^{R}$TC.

Another way to see the problem is to look at Carnap's assimilation of Ramsey's sentences (see Psillos 2000b; 2000c). Carnap noted that a theory TC is logically equivalent to the following conjunction: $^{R}$TC & ($^{R}$TC→TC), where the conditional $^{R}$TC→TC says that *if* there is some class of entities that satisfy the Ramsey-sentence, *then* the theoretical terms of the theory denote the members of this class. For Carnap, the Ramsey-sentence of the theory captured its factual content, and the conditional $^{R}$TC→TC captured its analytic content (it is a *meaning postulate*). This is so because the conditional $^{R}$TC→TC has no factual content: its own Ramsey-sentence, which would express its factual content if it had any, is logically true. As Winnie (1970, 294) observed, under the assumption that $^{R}$TC→TC – which is known as Carnap sentence – is a meaning postulate, it follows that $^{R}$TC↔TC, that is, that the theory is equivalent to its Ramsey-sentence.[13] In practice, this means that the Carnap sentence poses a certain restriction on the class of models that satisfy the theory: it excludes from it all models in which the Carnap-sentence fails. In particular, the models that are excluded are exactly those in which the Ramsey-sentence is true but the theory false. So if the Ramsey-sentence is true, the theory must be true: it cannot *fail* to be true. Is there a sense in which $^{R}$TC can be false? Of course, a Ramsey-sentence may be empirically inadequate. Then it is false. But if it *is* empirically adequate (i.e., if the structure of observable phenomena is embedded in one of its models), then it is bound to be true. For, as we have seen, given some cardinality constraints, it is guaranteed that there is an interpretation of the variables of $^{R}$TC in the theory's intended domain.

We can see why this result might not have bothered Carnap. If being empirically adequate is enough for a theory to be true, there is no extra issue of the truth of the theory to be reckoned with – apart of course from positing an extra domain of entities. Empiricism can thus accommodate the claim that theories are true, without going a lot beyond empirical adequacy.[14] Indeed, as I have argued elsewhere (Psillos 1999 Chapter 3; 2000b), Carnap took a step further. In his own account of Ramsey-sentences, he deflated the issue of the possible existential commitment to physical unobservable entities by taking the existentially bound Ramsey-variables to range over mathematical entities. Of the Ramsey-sentences he said:

> the observable events in the world are such that there are numbers, classes of such etc., which are correlated with the events in a prescribed way and which have among themselves certain relations; and this assertion is clearly a factual statement about the world. (Carnap 1963, 963)

Carnap's thought, surely, was not that the values of the variable are *literally* numbers, classes of them and so on. How possibly can a number be correlated with an observable event? Rather, his thought was that (a) the use of Ramsey-sentences does *not* commit someone to the existence of *physical* unobservable entities (let alone, to electrons in particular); and (b) the things that matter are the observable consequences of the Ramsey-sentence, its logical form, and its abstract claim of realisation.

Let me grant that this equation of truth with empirical adequacy is quite OK for an empiricist, though I should say in passing that it reduces much of science to nonsense and trivialises the search for truth.[15] But reducing truth to empirical adequacy *is* a problem for those who want to be realists, even if just about structure. For, it is no longer clear what has been left for someone to be realist about. Perhaps, the structural realist will insist that the range of the Ramsey-variables comprises *unobservable* entities and properties. It's not clear what the reason for this assertion is. What is it, in other words, that excludes all other interpretations of the range of Ramsey-variable?

Let's assume that, somehow, the range of Ramsey-variables is physical unobservable entities. It might be thought that, consistently with structuralism, the excess content of theories is given in the form of *non-formal* structural properties of the unobservables. Maxwell, for instance, didn't take all of the so-called structural properties to be purely formal (cf. Maxwell 1970b, 188). In Maxwell (1970a, 17), he took 'temporal succession, simultaneity, and causal connection' to be among the structural properties. His argument for this is hardly conclusive: 'for it is by virtue of them that the unobservables interact with one another and with observables and, thus, that Ramsey sentences have observable consequences.' Hearing this, Ramsey would have raised his eyebrow. In *Theories,* he noted: 'This causation is, of course, in the second system and must be laid out in the theory' (Ramsey 1931, 235).[16] The point, of course, is that we are in need of an independent argument for classifying some relations, for example, causation, as 'structural' and hence as knowable.

When it comes to causation, in particular, a number of issues need to be dealt with. First, what are its structural properties? Is causation irreflexive? Not if causation is persistence. Is causation asymmetric? Not if there is backward causation. Is causation transitive? Perhaps yes – but even this can be denied (in the case of probabilistic causation, for instance). Second, suppose that the structural properties of causation are irreflexivity, asymmetry and transitivity. If these properties constitute *all* that can be known of the relevant relation, what is there to distinguish causation from another relation with the same formal properties, for example, a relation of temporal ordering? Third, even if causation were a non-formal structural relation, why should it be the case that only its structural properties could be known?

Note a certain irony here. Suppose that since causation is a relation among events in the primary system, one assumes that it is the same relation that holds between unobservable events and between unobservable and observable events. This seems to be Maxwell's view. Causal knowledge in the primary system (i.e., causal knowledge concerning observables) is not purely structural. The (intrinsic) properties of events (or objects) by virtue of which they are causally related to one another are knowable. If causation is the very same relation irrespective of whether the relata are observable or unobservable, why should one assume that the (intrinsic) properties of unobservable events (or objects) by virtue of which they are causally related to one another are *not* knowable? There seems to be no ground for this asymmetry. In both cases, it may be argued, it is by virtue of the (intrinsic) properties that entities are causally related to each other. In both cases, it might be added, causal relations supervene on (or determined by) the intrinsic properties of observable or unobservable entities.[17] Indeed, these last points are fully consistent with Russell's (and Maxwell's) views. Recall that according to the causal theory of perception, which Maxwell also endorses, we (our percepts) are causally *affected* by the external objects (the stimuli, the causes). Now, this affection must be in virtue of these objects' intrinsic properties. (Surely, it is *not* by their formal properties.) The Helmholtz–Weyl principle (that different percepts are caused by different stimuli) implies that the different stimuli must differ in their *intrinsic* properties. So the latter are causally active and their causal activity is manifested in the different percepts they cause. In what sense then are they unknowable?[18]

The general point is that Maxwell's Ramsey-sentence approach to structuralism faces a dilemma. Either there is nothing left to be known except *formal* properties of the unobservable and observable properties or there are some knowable non-formal properties of the unobservable. In the former case, the Ramsey-sentence leaves it (almost) entirely open what we are talking about. In the latter case, we know a lot more about what the Ramsey-sentence refers to, but we thereby abandon pure structuralism.

## 9.6 Worrall and Zahar's structuralism

The points raised in the last section are particularly relevant to the Zahar–Worrall view (see Zahar 2001) that adopting the Ramsey-sentence of the theory is *enough* for being a realist about this theory. Naturally, they are aware of the problems raised so far. They do deal with Putnam's model-theoretic argument against realism and admit that if this argument is cogent, then the Ramsey-sentence of a(n) (epistemically ideal) theory is true. Note that a theory's being epistemically ideal includes its being empirically adequate. It's not hard to see that Putnam's argument is a version of the Newman problem.[19] Zahar (and his co-author, Worall, see Zahar 2001, 248) ask: 'should the structural realist be worried about these results?' And they note: '(...) the answer is decidedly negative'. So Zahar and Worrall do accept the equation of the truth of the theory with the truth of its Ramsey-sentence. In fact, they want to capitalise on this in order to claim that truth is achievable. They claim that two *seemingly* incompatible empirically adequate theories will have compatible Ramsey-sentences and hence they can both be true of the world (cf. ibid., 248–9).

We have already seen the price that needs to be paid for truth being achievable this way: truth is a priori ascertainable, given empirical adequacy and cardinality. But it is interesting to note that Zahar and Worrall are not entirely happy with this equation. They stress that 'the more demanding structural realist' can say a bit more. He (they?) can distinguish between different empirically adequate Ramsey-sentences using the usual theoretical virtues (simplicity, unification and so on), opt for the one that better exemplifies these virtues (e.g., it is more unified than the other) and claim that it is *this one* that should be taken 'to reflect – if only approximately – the real structure of W (the world)' (ibid., 249). I doubt that this, otherwise sensible, strategy will work in this case. For one, given that the theoretical virtues are meant to capture the explanatory power of a theory, it is not clear in what sense the Ramsey-sentence *explains* anything. If its truth is the same as its empirical adequacy, then the former cannot explain the latter. For another, there is something even more puzzling in the Zahar–Worrall claim. If two theories have compatible Ramsey-sentences, and if truth reduces to empirical adequacy, in what sense can the theoretical virtues help us deem one theory true and the other false? There is no such sense, I content. There *could* be, if truth and empirical adequacy were distinct. But this is exactly what the Zahar–Worrall line denies. Could they simply say that there is a sense in which one Ramsey-sentence is *truer* than the other? They could, but only if the truth of the theory answered to something different from its empirical adequacy. If, for instance, it was claimed that a theory is true if, on top of its being empirically adequate, it captures the natural structure of the world, it becomes clear that (a) one theory could be empirically adequate and yet false; and (b) one of two empirically adequate theories could be truer than the other.[20]

In his reply to Russell's structuralism, Newman pointed to a way in which Russell's claim would *not* be trivial, namely, if the relation that generated the required structure *W* was 'definite', that is if we knew (or claimed that we knew) *more* about what it is than that it exists and has certain formal properties. Couldn't we distinguish between 'important' and 'unimportant' relations and stay within structuralism? Not really. The whole point is precisely that the notion of 'important relation' cannot admit of a purely structuralist understanding. Newman saw this very clearly (see Newman 1928, 147). In order to pick as important one among the many relations which generate the same structure on a domain, we have to go beyond structure and talk about *what* these relations are, and *why* some of them are more important than others.

It's not hard to see that the very same objection can be raised against a Zahar–Worrall structural realism. And it is equally obvious what the remedy could be. Structural realists should have a richer understanding of the relations that structure the world. Suppose there is indeed some definite relation (or a network, thereof) that generates the structure of the world. If this is the case, the claim that the structure $W$ of the physical world is isomorphic to the structure $W'$ that satisfies an empirically adequate Ramsey-sentence would be far from trivial. It would require, and follow from, a comparison of the structures of two independently given relations, say $R$ and $R'$. But structural realists as well as Russell *deny* any independent characterisation of the relation $R$ that generates the structure of the physical world. On the contrary, structural realists and Russell insist that we can get at this relation $R$ only by knowing the structure of another relation $R'$, which is isomorphic to $R$. We saw that the existence of $R$ (and hence of $W$) follows logically from some fact about cardinality.

Treating these relations as 'definite' would amount to an abandonment (or a strong modification) of structuralism.[21] The suggestion here is that among all those relations-in-extension which generate the same structure, only those which express *real relations* should be considered. But specifying which relations are real requires knowing something *beyond* structure, namely, which extensions are 'natural', that is, which subsets of the power set of the domain of discourse correspond to natural properties and relations. Having specified these natural relations, one may abstract away their content and study their structure. But if one begins with the structure, one is in no position to tell *which* relations one studies and *whether* they are natural or not.

## 9.7 Ramsey and Newman's problem

As noted above, Ramsey's crucial observation was that the excess content of the theory is seen when the theory is formulated as expressing an existential judgement. If, on top of that, Ramsey meant to assert something akin to the structural realist position, that is, that this excess content, so far as it is knowable, is purely *structural*, he would have landed squarely on the Newman problem. So should this view be attributed to Ramsey?

Before I canvass a negative answer, let me go back to Russell once more. Russell (1927) took theories to be hypothetico-deductive systems and raised the issue of their interpretation. Among the many 'different sets of objects (that) are abstractly available as fulfilling the hypotheses', he distinguished those that offer an 'important' interpretation (ibid., 4–5), that is an interpretation which connects the theory (as an abstract logico-mathematical system) to the empirical world. This was important, he thought, because all the evidence there is for physics comes from perceptions. He then went on to raise the central question that was meant to occupy the body of his book: when are physical theories *true*? As Russell (1927, 8–9) put it, there is a wider sense in which physics is true:

> Given physics as a deductive system, derived from certain hypotheses as to undefined terms, do there exist particulars, or logical structures composed of particulars, which satisfy these hypotheses?

'If', he added, 'the answer is in the affirmative, then physics is completely "true"'. He took it that his subsequent structuralist account, based on the causal theory of perception, was meant to answer the above question affirmatively. Russell's view has an obvious similarity to Ramsey's: theories as hypothetico-deductive structures should be committed to an existential claim that *there is* an interpretation of them. But there is an interesting dissimilarity between Russell and Ramsey. Russell thinks that *some* interpretation is important (or more important than others), whereas Ramsey has no place for this in his view. Russell might well identify the theory with a *definite description*: there is a unique (important) interpretation such that the axioms of the theory are true of it. But, as we have seen, one of Ramsey's insights is that there is no reason to think of theories as *definite* descriptions – that is as requiring uniqueness.

It seems that it was this Russellian question that inspired Ramsey to formulate his own view of theories as existential judgements. In fact, there is some evidence for it. In a note on Russell's *The Analysis of Matter*, Ramsey (1991, 251) says:

$$\text{Physics says} = \text{is true if } (\exists\, \alpha, \beta, \ldots \mathrm{R}, \mathrm{S}) : \mathrm{F}(\alpha, \beta, \ldots \mathrm{R}, \mathrm{S} \ldots) \qquad (1)$$

He appended a reference – 'Russell p. 8' – to *The Analysis of Matter*.[22]

Unlike Russell, Ramsey did *not* adopt a structuralist view of the content of theories. This may be evinced by what he goes on to say: the propositional functions $\alpha$ and $R$ should be 'nonformal'. And he adds: 'Further $F$ must not be tautological as it is on Eddington's view'. As it is clear from another note (1991, 246–50), Ramsey refers to Arthur Eddington's *The Nature of the Physical World*. In his review of this book, Braithwaite criticised Eddington severely for trying to turn physics 'from an inductive science to a branch of mathematics' (Braithwaite 1929, 427). According to Braithwaite, Eddington tried to show how the laws of physics reduce to mathematical identities, which are derivable from very general mathematical assumptions. This must be wrong, Braithwaite thought, for in mathematics 'we never know what we are talking about', whereas in physics 'we do know (or assume we know) something of what we are talking about – that the relata have certain properties and relations – without which knowledge we should have no reason for asserting the field laws (even without reference to observed quantities' (ibid., 428). The point might not be as clear as it ought to have been, but, in effect, Braithwaite argued against Eddington that natural science would be trivialised if it was taken to aim at achieving only knowledge of structure.[23]

I don't know whether Ramsey discussed Eddington's book with Braithwaite or whether he had read Braithwaite's review of it (though he *had* read Eddington's book – see 1991, 246–50). It is nonetheless plausible to say that he shared Braithwaite's view when he said of the relation *F* that generates the structure of a theory that it should not be tautological 'as it is on Eddington's view'. In fact, in the very same note, Ramsey claims that in order to fix some interpretation of the theory we need 'some restrictions on the interpretation of the other variables. That is, all we know about β, S is not that they satisfy (1)'.

So I don't think Ramsey thought that viewing theories as existential judgements entailed that only structure (plus propositions of the primary system) could be known. It's plausible to argue that Ramsey took Ramsey-sentences (in *his* sense) to require the existence of *definite* relations, whose nature might not be fully determined, but which is nonetheless constrained by some theoretical and observational properties. To judge the plausibility of this interpretation, let's look into some of his other papers.

In *The Foundations of Mathematics*, Ramsey insisted on the distinction between classes and relations-in-extension, on the one hand, and real or actual properties and relations, on the other. The former are identified extensionally, either as classes of objects or as ordered n-tuples of objects. The latter are identified by means of predicates. Ramsey agreed that an extensional understanding of classes and relations is necessary for mathematics. Take, for instance, Cantor's concept of class-similarity. Two classes are similar (i.e., they have the same cardinality) iff there is an one–one correspondence (relation) between their domains. This relation, Ramsey (1931, 15) says, is a relation-in-extension: there needn't be any actual (or real) relation correlating the two classes. The class of male angels may have the same cardinality with the class of female angels, so that the two classes can be paired off completely, without there being some real relation ('such as marriage') correlating them (ibid., 23). But this is not all there is to relations. For it may well be the case that two classes have the same cardinality because there is a 'real relation or function $f(x, y)$ correlating them term by term' (ibid.) He took it that the real propositional functions are determined 'by a description of their senses or imports' (ibid., 37). In fact, he thought that appealing to the meaning of propositional functions is particularly important when we want to talk of functions of functions (Ramsey's $f(\phi x)$), that is (higher-level) propositional functions $f$ whose values are other propositional functions ($\phi x$). He wrote: 'The problem is ultimately to fix as values of $f(\phi x)$ some definite set of propositions so that we can assert their logical sum or product' (ibid., 37). And he took it that the best way to determine the range of the values of $f(\phi x)$ is to appeal to the meanings of the lower-level propositional functions ($\phi x$) (ibid., 36–7).

Recall Ramsey's Ramsey-sentence *(*$\exists\, \alpha, \beta, \gamma$*) : dictionary · axioms*. The open formula *dictionary · axioms (*$\alpha, \beta, \gamma$*)* is a higher-level propositional function, whereas the values of $\alpha$, $\beta$, $\gamma$ are lower-level propositional functions. The Ramsey-sentence itself expresses the logical sum of the propositions that result when specific values are given to $\alpha$, $\beta$, $\gamma$. The situation is exactly analogous to the one discussed by Ramsey above. So, it's plausible to think that the values of $\alpha$, $\beta$, $\gamma$ are some definite properties and relations. That is, they are not *any* class or relation-in-extension that can be defined on the domain of discourse of the Ramsey-sentence.

This point can be reinforced if we look at Ramsey's *Universals*. Among other things, Ramsey argues that the extensional character of mathematics 'is responsible for that great muddle the theory of universals', because it has tended to obscure the important distinction between those propositional functions that are names and those that are incomplete symbols (cf. ibid., 130–1 & 134). The mathematical logician is interested only in classes and relations-in-extension. The difference between names and incomplete symbols won't be reflected in any difference in the classes they define. So the mathematician disregards this difference, though, as Ramsey says, it is 'all important to philosophy' (ibid., 131). The fact that some functions cannot stand alone (i.e., they are incomplete symbols) does not mean that 'all cannot' (ibid.) Ramsey takes it that propositional functions that are names might well name 'qualities' of individuals (cf. ibid., 132). Ramsey puts this idea to use in his famous argument that there is no difference between particulars and universals.[24] But the point relevant to our discussion is that propositional functions can be names.

Given (a) Ramsey's view that the propositional functions of physics should be non-formal, (b) his insistence on real or actual properties and relations and (c) his view that at least *some* relations can be named by propositional functions, it seems plausible to think that he took the variables of *his* Ramsey-sentence to range over *real properties and relations* – some of which could be named. I am not aware of a passage in his writings which says *explicitly* that the variables of the Ramsey-sentence range over real or actual properties and relations. But his contrasting of mathematics (in which the variables are purely extensional) to science suggests that he might well have taken the view ascribed to him above. Now, the other claim, namely, that some of the Ramsey-variables can be names also follows from his view, seen in Section 9.3, that some propositional functions can give way to *names* of properties, as science grows.

If I am right, the Newman problem cannot be raised against Ramsey's views. Ramsey takes theories to imply the existence of definite (or real) relations and properties. Hence, it's no longer trivial (in the sense explained above) that if the theory is empirically adequate, it is true. His Ramsey-sentences can be seen as saying that there are *real properties and relations* such that.... Note that, in line with Ramsey's denial of a distinction between universals and particulars, the existentially bound variables should be taken to quantify over properties and relations in a metaphysically non-committal way: they quantify over properties and relations which are not universals in the traditional sense, which renders them fundamentally different from particulars.[25]

The corollary is that Ramsey's views cannot be described as *pure* structuralism. The claim that there are real properties and relations is not structural because it specifies the *types* of structure that one is interested in. Besides, Ramsey does not claim that only the structure (or the structural properties) of these relations can be known. Well, it might. Or it might not. This is certainly a contingent matter.

If my interpretation is right, I have a hurdle to jump. It comes from Ramsey's comment on 'the best way to write our theory'. He says: 'Here it is evident that $\alpha$, $\beta$, $\gamma$ are to be taken purely extensionally. Their extensions may be filled with intensions or not, but this is irrelevant to what can be deduced in the primary system' (Ramsey 1931, 231). This comment, however, is consistent with my reading of his views. The propositional variables may range over real properties and relations, but when it comes to what can be *deduced* in the primary system what matters is that they are of a certain logical type, which the Ramsey-sentence preserves anyway. Indeed, deduction cuts through content and that's why it is important. In any case, the comment above would block my interpretation only if what really mattered for theories was what could be deduced in the primary system. I have already said enough, I hope, to suggest that this view was *not* Ramsey's.

## 9.8 Ramseyan humility

Let me end by sketching an image of scientific theories to which the above interpretation of Ramsey's Ramsey-sentences might conform. As already noted, I call it *Ramseyan humility*.

We treat our theory of the world as a *growing existential statement*. We do that because we want our theory to express a judgement: to be truth-valuable. In writing the theory we commit ourselves to the existence of things that make our theory true and, in particular, to the existence of unobservable things that cause or explain the observable phenomena. We don't *have to* do this. But we think we are better off doing it, for theoretical, methodological and practical reasons. So we are *bold*. Our boldness extends a bit more. We take the world to have a certain structure (to have *natural joints*). We have independent reasons to think of it, but in any case, we want to make our theory's claim to truth or falsity substantive. The theoretical superstructure of our theory is not just an idle wheel. We don't want our theory to be true just in case it is empirically adequate. We want the structure of the world to act as an *external constraint* on the truth or falsity of our theory. So we posit the existence of a natural structure of the world (with its natural properties and relations). We come to realise that this move is *not* optional once we made the first bold step of positing a domain of unobservable entities. These entities are powerless without properties and relations and the substantive truth of our theories requires that these are real (or natural) properties and relations.[26]

That's, more or less, where our boldness ends. We don't want to push our (epistemic) luck hard. We want to be *humble* too. We don't foreclose the possibility that our theory might not be uniquely realised. So we don't require uniqueness: we don't turn our growing existential statement into a definite description. In a sense, if we did, we would no longer consider it as growing. We allow a certain amount of indeterminacy and hope that it will narrow down as we progress. Equally, we don't foreclose the possibility that what the things we posited are might not be found out. Some things must exist if our theory is to be true and these things must have a natural structure if this truth is substantive. Humility teaches us that there are many ways in which these commitments can be spelt out. It also teaches us that, in the end, we might not be lucky. We don't, however, draw a sharp and principled distinction between what can be known and what cannot. We are not lured into thinking that only the structure of the unobservable world can be known, or that only the structural properties of the entities we posited are knowable, or that we are cognitively shut off from their intrinsic properties. We are reflective beings after all, and realise that these claims need independent argument to be plausible. We read Kant, we read Russell, Schlick, Maxwell and Redhead, but we have not yet been persuaded that there is a sound independent argument. So we choose to be open-minded about this issue. The sole arbiter we admit is our give-and-take with the world.

A sign of our humility is that we treat what appear to be *names* of theoretical entities as variables. We refer to their values indefinitely, but we are committed to their being *some* values that make the theory true. As science grows, and as we acquire some knowledge of the furniture of the world, we modify our growing existential statement. We are free to replace a variable with a name. We are free to add some new findings within our growing existential statement. We thereby *change* our theory of the world, but we had anticipated this need. That's why we wrote the theory as a *growing* existential statement. We can bring under the same roof continuity and change. The continuity is secured by the bound Ramsey-variables and the change is accommodated by adding or deleting things within their scope.

In the meantime, we can accommodate substantial disagreement of two sorts. Scientific disagreement: what exactly are the entities posited? In fostering this kind of disagreement, we are still able to use the theory to draw testable predictions about the observable world. But we do not thereby treat the theoretical part of the theory simply as an aid to prediction. We have not conceded that all that can possibly be known of the entities posited is that they exist. We can also accommodate metaphysical disagreement: what is the metaphysical *status* of the entities posited? Are they classes? Universals? Tropes? Some kind of entity which is neutral? Still, in fostering this kind of disagreement, we have taken a metaphysical stance: whatever else these entities are, they should be natural.

To be conciliatory, I could describe Ramseyan humility as *modified structuralism*. Structuralism emerges as a humble philosophical thesis, which rests on a bold assumption – without which it verges on vacuity – namely, that the world has a natural structure that acts as an external constraint on the truth or falsity of theories. I don't claim that the image sketched *is* Ramsey's. But he might have liked it. In any case, I take it that something like it is true. It's not attractive to someone who is not, however, a realist of some sort. But it is flexible enough to accommodate realisms of all sorts.

# Part III

# Inference to the Best Explanation

# 10
# Simply the Best: A Case for Abduction

In this chapter I will do two things. First, I shall formulate what I think is *the basic problem* of any attempt to characterise the abstract structure of scientific method, namely, that it has to satisfy two conflicting desiderata: it should be ampliative (content-increasing) and it should confer epistemic warrant on its outcomes (cf. Gower 1998; Psillos 1999). Second, and after I have examined two extreme solutions to the problem of the method, namely, Enumerative Induction (EI) and the Method of Hypothesis, I will try to show that abduction, suitably understood as inference to the best explanation (IBE), offers the best description of scientific method and solves the foregoing problem in the best way: it strikes the best balance between ampliation and epistemic warrant.

The general framework I will follow is Pollock's (1986) analysis of defeasible reasoning in terms of the presence or absence of *defeaters*. This framework makes possible the investigation of the conditions under which defeasible reasoning can issue in warranted beliefs. I shall also raise and try to answer some general philosophical questions concerning the epistemic status of abduction.

In what follows, I shall deliberately leave aside all the substantive issues about the nature of explanation.[1] This is partly because they are just too many and messy to be dealt with in this chapter and partly because I think that – barring some general platitudes about the nature of explanation – my claims about IBE should be neutral vis-à-vis the main theories of explanation.[2] At any rate, I think that the very possibility of IBE as a warranted ampliative method must be examined independently of specific models of the explanatory relationship between hypotheses and evidence. Ideally, IBE should be able to accommodate different conceptions of explanation. This last thought implies that abduction (i.e., IBE) is not usefully seen as a species of ampliative reasoning, but rather as a *genus* whose several species are distinguished by plugging assorted conceptions of explanation in the reasoning schema that constitutes the genus. For instance, if the relevant notion of explanation is revealing of causes, IBE becomes an inference to the best causal explanation. Or, if the relevant notion of explanation is subsumption under laws, IBE becomes a kind of inference to the best Deductive-Nomological explanation, and so forth. Given that there is too much disagreement on the notion of explanation, and given that no account offered in the literature so far seems to cover fully all aspects of explanation, it seems to me methodologically useful to treat the reference to explanation in IBE as a 'placeholder' which can be spelled out in different ways in different contexts. Some philosophers may think that this approach to IBE renders it an unnatural agglomeration of many different types of reasoning where explanatory considerations are involved. I think it is at least premature to call this agglomeration 'unnatural'. After all, as I hope to show in this chapter, the general ways in which explanatory considerations can enter into defeasible reasoning can be specified without a prior commitment to the nature of the explanatory relation.

## 10.1 Ampliation and epistemic warrant

Any attempt to characterise the abstract structure of scientific method should make the method satisfy two general and intuitively compelling desiderata: it should be ampliative *and* epistemically probative. Ampliation is necessary if the method is to deliver informative hypotheses and theories, namely, hypotheses and theories that exceed in content the observations, data, experimental results and, in general, the experiences which prompt them. This 'content-increasing' aspect of scientific method is indispensable, if science is seen, at least *prima facie*, as an activity which purports to extend our knowledge (and our understanding) beyond what is observed by means of the senses. But this ampliation would be merely illusory, *qua* increase of *content*, if the method was not epistemically probative: if it did not convey epistemic warrant to the excess content produced thus (viz., hypotheses and theories). To say that the method produces – as its output – more information than what there is in its input is one thing. To say that this extra information can reasonably be held to be warranted is quite another. Now, the real problem of the scientific method is that these two plausible desiderata are not jointly satisfiable. Or, to weaken the claim a bit, the problem is that there seems to be good reason to think that they are not jointly satisfiable. The tension between them arises from the fact that ampliation does not carry its epistemically probative character on its sleeves. When ampliation takes place, the output of the method can be false while its input is true. The following question then arises: what makes it the case that the method conveys epistemic warrant to the intended output rather than to any other output which is consistent with the input? Notice that ampliation has precisely the features that deduction lacks. Suppose one thought that a purely deductive method is epistemically probative in the following (conditional) sense: if the input (premises) is warranted, then the method guarantees that the output cannot be less warranted than the input. No ampliative method can be epistemically probative in the above sense. But can there be any other way in which a method can be epistemically probative? If the method is not such that the input excludes all but one output, in what sense does it confer any warrant on a certain output?

'In no sense', is the strong sceptical (Humean) answer. The sceptic points out that any attempt to strike a balance between ampliation and epistemic warrant is futile for the following reason. Given that ampliative methods will fail to satisfy the aforementioned conditional, they will have to base any differential epistemic treatment of outputs which are consistent with the input on some *substantive and contingent assumptions* (e.g., that the world has a natural-kind structure, or that the world is governed by universal regularities, or that observable phenomena have unobservable causes, etc.) It is these substantive assumptions that will do all the work in conferring epistemic warrant on some output rather than another. But, the sceptic goes on, what else, other than ampliative reasoning itself, can possibly establish that these substantive and contingent assumptions are true of the world? Arguing in a circle, the sceptic notes, is inevitable and this simply means, he concludes, that the alleged balance between ampliation and epistemic warrant carries no rational compulsion with it. In other words, the sceptic capitalises on the fact that in a purely deductive (non-ampliative) method, the transference of the epistemic warrant from the premises to the conclusion is parasitic on their formal (deductive) relationship, whereas in an ampliative method the alleged transference of the epistemic warrant from the premises to the conclusion depends on substantive (and hence challengeable) background beliefs and considerations.[3]

A standard answer to the problem of method is to grant that the sceptic wins. I think this is too quick. Note that the sceptical challenge is far from intuitively compelling. It itself relies on a substantive *epistemic* assumption: that any defence of an ampliative but epistemically probative method should simply mirror some formal relations between the input and the output of the method and should depend on no substantive and contingent assumptions whose truth cannot be established by independent means. This very assumption is itself subject to criticism.[4] First, if it is accepted, it becomes a priori true that there can be no epistemically probative ampliative method. Yet, it may be reasonably argued that the issue of whether or not there can be ampliative yet epistemically probative methods should hinge on information about the actual world and its structure (or, also on information about those possible worlds which have the same nomological structure as the actual).

A proof that a method could be both ampliative and epistemically probative in *all* possible worlds (a proof which we have *no* reasons to believe is forthcoming) would certainly show that it can have these features in the actual world. But the very request of such a proof (one that could persuade the sceptic) relies on the substantive assumption that an epistemically probative method should be totally insensitive to the actual features (or structure) of the world. This request is far from compelling. We primarily need our methods to be the right ones for the world we live in. If the range of their effectiveness is larger, that's a pleasant bonus. We can certainly live without it. Second, if the sceptical assumption is accepted, even the possibility of epistemically probative demonstrative reasoning becomes dubious. Truth-transmission, even though it is guaranteed by deductive reasoning, requires some truths to start with. The truth of any substantive claims that feature in the premises of a deductive argument can only be established by ampliative reasoning, and hence it is equally open to the sceptical challenge.[5] The point here is not that relations of deductive entailment between some premise *P* and a conclusion *Q* fail to offer an epistemic warrant for accepting *Q*, *if one already warrantedly accepts P*. Rather, the point is that coming to accept as true a premise *P* with any serious content will typically involve some ampliative reasoning. The sceptical challenge is not incoherent. But if its central assumption is taken seriously, what is endangered is not just the very possibility of any kind of learning from experience, but also any kind of substantive reasoning.

There is, however, something important in a modest reading of the sceptical answer to the problem of method: if we see it as a challenge to offer a satisfactory account of method which is both ampliative and epistemically probative, we can at least make some progress in our attempt to understand under what conditions (and under what substantive assumptions) the two desiderata can co-exist.

## 10.2 Between two extremes

To start making this progress, we need to see how the two standard accounts of scientific method fare vis-à-vis the two desiderata. We'll look at Enumerative Induction (EI) and crude hypothetico-deductivism (HD) (or, the method of hypothesis) and compare them in terms of the strength of ampliation and the strength of the epistemic warrant. However, let me first make an important note.

### 10.2.1 Defeasibility and defeaters

The very idea of ampliation implies that the outcome of the application of an ampliative method (or of a mode of ampliative reasoning) can be defeated by new information or evidence. Unlike deductive methods, ampliative methods are *defeasible*. The issue here is not just that further information can make the output not to logically follow from the input. It is that further information can remove the *warrant* for holding the output of the method. Further information can make the previous input not be strong enough to warrant the output. Following Pollock (1986 Chapter 2, Section 3), we can call '*prima facie*' or 'defeasible' any type of reason which is not conclusive (in the sense that it is not deductively linked with the output it is a reason for). Given that ampliative reasoning is defeasible, we can say that such reasoning provides *prima facie warrant* for an output (belief). What Pollock has rightly stressed is that to call a warrant (or a reason) *prima facie* is not to degrade it, *qua* warrant or reason. Rather, it is to stress that (a) it can be defeated by further reasons (or information); and (b) its strength, *qua* reason, is a function of the presence or absence of 'defeaters'. 'Defeaters' are the factors (generally, reasons or information) that, when taken into account, they can remove the *prima facie* warrant for an outcome (belief). On Pollock's insightful analysis of reasoning and warrant, the presence or absence of defeaters has directly to do with the degree in which one is warranted to hold a certain belief. Suppose that a subject *S* has a *prima facie* (nonconclusive) reason *R* to believe *Q*. Then *S* is warranted to believe that *Q* on the basis of *R*, *unless* either there are further reasons *R'* such that, were they to be taken into account, they would lead *S* to doubt the integrity of *R* as a reason for *Q*, or there are strong (independent) reasons to hold not-*Q*. Generalising this idea to the problem of method, we may say that the presence or absence of defeaters is directly linked to the degree in which an ampliative method can confer epistemic warrant on an outcome, that is, the degree in which it can be epistemically probative. To say that *S* is *prima facie* warranted to accept the outcome *Q* of an ampliative method is to say that although it is possible that there are defeaters of the outcome Q, such defeaters are not actual. In particular, it is to say that *S* has considered several possible defeaters of the reasons offered for this outcome Q and has shown that they are not present. If this is done, we can say that there are no *specific* doubts about the outcome of the method and, that belief in this outcome is *prima facie* warranted.

This talk of defeaters is not abstract. There are general *types* of defeater that one can consider. Hence, when it comes to considering whether an outcome is warranted, there are certain things to look at such that, if present, would remove the warrant for the outcome. Even if it is logically possible that there could be considerations that would undercut the warrant for the outcome (a possibility that follows from the very idea of defeasibility), the concrete issue is whether or not there actually are such considerations (actual defeaters).[6] Besides, if the reasoner has done whatever she can to ensure that such defeaters are not present in a particular case, there is a strong sense in which she has done what it can plausibly be demanded of her in order to be epistemically justified. Pollock (1986, 38–9) has identified two general types of defeater: 'rebutting' and 'undercutting'. Suppose, for simplicity, that the ampliative method offers some *prima facie* reason *P* for the outcome Q. A factor *R* is called a rebutting defeater for *P* as a reason for Q if and only if *R* is a reason for believing not-Q. And a factor *R* is called an undercutting defeater for *P* as a reason for Q if and only if *R* is a reason for denying that *P* offers warrant for Q.[7] So, considering whether or not Q is warranted on the basis of *P* one has to consider whether or not there are rebutting and undercutting defeaters. Taking all this into account, let us look at the two extreme cases of ampliative method.

### 10.2.2 Enumerative induction

EI is based on the following: if one has observed *n* As being B and *no* As being not-B, and if the evidence is enough and variable, one should infer that (with probability) 'All As are B'. The crux of EI is that ampliation is effected by *generalisation*. We observe a pattern among the data (or, among the instances of two attributes), and then generalise it so that it covers all the values of the relevant variables (or all instances of the two attributes). For obvious reasons, we can call EI, the 'more-of-the-same' method (cf. Lipton 1991, 16). The prime advantage of EI is that it is content-increasing in a, so to speak, 'horizontal way': it allows the acceptance of generalisations based on evidence in a way that stays close to what has been actually observed. In particular, no new entities (other than those referred to in (descriptions of) the data) are introduced by the ampliation. Let me call this *minimal ampliation*. The basic substantive assumptions involved in this ampliation are that (a) there are projectable regularities among the data; and (b) the pattern detected among the data (or the observations) in the sample is representative of the pattern (regularity) in the whole relevant population. The *prima facie* warrant that EI confers on its outcomes is based on these substantive assumptions. But this warrant – and the assumptions themselves – are subject to evaluation. EI admits of both undercutting and rebutting defeaters. If there are specific reasons to doubt that the pattern among the data can be projected to a lawful regularity in the population, the projection is not warranted.[8] If there are specific reasons to doubt the fairness of the sample, the projection is no longer warranted. Note that although the sample may be unfair (e.g., the sample might involve only ravens in a certain region), the conclusion (viz., that all ravens are black) may well be true. Yet, knowing that the sample was unfair does remove the warrant for the conclusion. These are cases of undercutting defeaters. Besides, EI admits of rebutting defeaters. If we find a negative instance (e.g., a black swan) the warrant for the conclusion (e.g., that all swans are white) is completely removed.[9] In EI we know precisely what kind of defeaters can remove the *prima facie* warrant for making the ampliation (generalisation). And, on very many occasions, we (a) can certify the presence or absence of defeaters; and (b) can withhold the conclusion until we have strong reasons to believe that the potential defeaters are not present (e.g., by making meticulous search for cases which would rebut the conclusion). Given the very specific character of defeaters in EI, and the general feasibility of the search for defeaters, we can say that EI can be *maximally epistemically probative* (among ampliative methods). Here again, the point is not that the sceptic loses. Nor is it that EI *is* maximally epistemically probative. Rather, the point is that if ampliative – and hence defeasible – methods can be warranted at all based on the presence or absence of defeaters, and given that in the case of EI we know exactly what defeaters we should look for and how to do it, EI fares best in terms of how warranted an outcome of a successful (undefeated) application of EI can be.

EI is minimally ampliative and maximally epistemically probative. But this is precisely the problem with EI: that what we gain in (epistemic) austerity we lose in strength (of ampliation). EI is too restrictive. It cannot possibly yield any hypothesis about the causes of the phenomena. Nor can it introduce new entities. The basic problem is that the input and the output of EI are couched in the same vocabulary: conclusions that state generalisations are necessarily couched in the vocabulary of the premises. Hence, EI cannot legitimately introduce new vocabulary. Hence, it cannot possibly be used to form ampliative hypotheses that refer to entities whose descriptions go beyond the expressive power of the premises.[10]

### 10.2.3 The method of hypothesis

Let us turn to the crude version of the 'method of hypothesis' (HD). This is based on the following: Form a hypothesis H and derive some observational consequences from it. If the consequences are borne out, the hypothesis is confirmed (accepted). If they are not borne out, the hypothesis is disconfirmed (rejected). The crux of the method is that a hypothesis is warrantedly accepted on the basis of the fact that it entails all available relevant evidence. In HD, ampliation is effected by *confirmation*. An ampliative hypothesis H is accepted because it gets confirmed by the relevant evidence. To be sure, the operation of HD is more complicated. The observational consequences follow from the conjunction of H with some statements of initial conditions, other auxiliary assumptions and some bridge-principles which connect the vocabulary in which H is couched and the vocabulary in which the observational consequences are couched. It is this bridge-principles that make HD quite powerful, since they allow for what I shall call 'vertical extrapolation' – to be contrasted with the 'horizontal extrapolation' characteristic of EI. The content of H may well be much richer than the content of the relevant observational consequences and the deductive link between the two contents is guaranteed by the presence of bridge-principles.

The prime attraction of HD is precisely that is can be content-increasing in a, so to speak, vertical way: it allows the acceptance of hypotheses about the, typically unobservable, causes of the phenomena. In particular, new entities (other than those referred to in the data) are introduced by the ampliation. In contrast to EI, let me call this *maximal ampliation*. The basic substantive assumptions involved in this type of ampliation are that (a) there are causally and explanatory relevant entities and regularities behind the observed data or phenomena; and (b) the pattern detected among the data (or the observations) is the causal-nomological outcome of entities and processes behind the phenomena. What about the warrant that HD confers on its outcomes? As in the case of EI, we should look at the possible defeaters of the reasons offered by HD for the acceptance of a hypothesis H. The rebutting defeaters seem to be clear-cut: if the predicted observation is not borne out, then – by modus tollens – the hypothesis is refuted. This seems quite compelling, yet there are well-known problems. As we have just seen, it is typically the case that, in applications of HD, the predictions follow from the conjunction of the hypothesis with other auxiliary assumptions and initial and boundary conditions. Hence, when the prediction is *not* borne out, it is the whole cluster of premises that gets refuted. But HD alone cannot tell us how to apportion praise and blame among them. At least one of them is false but the culprit is not specified by HD. It might be that the hypothesis is wrong, or some of the auxiliaries were inappropriate. A possible rebutting defeater (a negative prediction) does not carry with it the epistemic force to defeat the hypothesis and hence to remove the warrant for it. (This is a version of the Duhem-Quine problem.) In order for the rebutting defeater to do its job, we need further information, namely, whether the hypothesis is warranted enough to be held on, or whether the auxiliaries are vulnerable to substantive criticism and so on. All these considerations go a lot beyond the deductive link between hypotheses and data that forms the backbone of HD and are not incorporated by the logical structure of HD.

What about the undercutting defeaters? It's not clear what these are. It seems a good idea to say that an undercutting defeater for a hypothesis H which does conform to the observations is another hypothesis H* which also conforms to the observations. For if we know that there is another H*, it seems that our confidence about H is negatively affected. The *prima facie* warrant for H (based as it is on the fact that H entails the evidence) may not be totally removed, but our confidence that H is correct will surely be undermined. To put the same point in a different way, if our warrant for H is based *solely* on the fact that it entails the evidence, insofar as there is another hypothesis H* which also entails the evidence, H and H* will be equally warranted. It may be that H* entails H, which means that, on probabilistic considerations, H will be at least as probable as H*. But this is a special case. The general case is that H and the alternative hypothesis H* will be mutually inconsistent. Hence, HD will offer no way to discriminate between them in terms of warrant. The existence of each alternative hypothesis will act as an undercutting defeater for the rest of them. Given that, typically, for any H there will be alternative hypotheses which also entail the evidence, HD suffers from the existence of just too many undercutting defeaters. All this can naturally lead us to the conclusion that HD is *minimally epistemically probative*, since it does not have the resources to show how the undercutting defeaters can be removed.[11]

HD is maximally ampliative and minimally epistemically probative. But this is precisely the problem with it: that what we gain in strength (of ampliation) we lose in (epistemic) austerity. Unlike EI, it can lead to hypotheses about the causes of the phenomena. And it can introduce new entities: it can also be 'vertically ampliative'. But, also unlike EI, HD is epistemically too permissive. Since there are, typically, more than one (mutually incompatible) hypothesis which entail the very same evidence, if a crude 'method of hypothesis' were to license any of them as probably true, it would also have to license all of them as probably true. This permissiveness leads to absurdities. The crude 'method of hypothesis' simply lacks the discriminatory power that scientific method ought to have.[12]

## 10.3 A case for abduction

Faced with these two extreme solutions to the problem of the scientific method, the question is whether there can be a characterisation of the method that somehow moves in-between them. So far, we have noted that ampliation is inversely proportional to epistemic warrant. This is clearly not accidental, since ampliation amounts to risk and the more the risk taken, the less the epistemic security it enjoys. But it is an open issue whether or not there can be a way to strike a balance between ampliation and epistemic warrant, or (equivalently) between strength and austerity. In particular, it is an open issue whether there can be a characterisation of the method which strikes a balance between EI's restrictive ampliation and HD's epistemic permissiveness. I want to explore the suggestion that abduction, if suitably understood as inference to the best explanation (IBE), can offer the required trade-off. But first, what is abduction?

### 10.3.1 What is abduction?

I am going to leave aside any attempt to connect what follows with Peirce's views on abduction.[13] Rather, I shall take Gilbert Harman's (1965) as the *locus classicus* of the characterisation of IBE. 'In making this inference', Harman (1965, 89) notes,

> one infers, from the fact that a certain hypothesis would explain the evidence, to the truth of that hypothesis. In general, there will be several hypotheses that might explain the evidence, so one must be able to reject all such alternative hypotheses before one is warranted in making the inference. Thus one infers, from the premise that a

given hypothesis would provide a 'better' explanation for the evidence than would any other hypothesis, to the conclusion that the given hypothesis is true.

Following John Josephson (1994, 5), IBE can be put schematically thus:

(A)
D is a collection of data (facts, observations, givens).
H explains D (H would, if true, explain D)
No other hypothesis can explain D as well as H does.

Therefore, H is probably true.[14]

It is important to keep in mind that, on IBE, it is not just the semantic relation between the hypothesis and the evidence which constitutes the *prima facie* warrant for the acceptance of the hypothesis. Rather, it is the *explanatory quality* of this hypothesis, on its own but also taken in comparison to others, which contributes essentially to the warrant for its acceptability. What we should be after here is a kind of measure of the explanatory power of a hypothesis. Explanatory power is connected with the basic function of an explanation, namely, to provide understanding. Whatever the formal details of an explanation, it should be such that it enhances our understanding of why the explanandum happened. This can be effected by incorporating the explanandum into the rest of our background knowledge by providing some link between the explanandum and other hypotheses that are part of our background knowledge. Intuitively, there can be better and worse ways to achieve this incorporation – and hence the concomitant understanding of the explanandum. For instance, an explanation which does not introduce gratuitous hypotheses in its story, or one that tallies better with the relevant background knowledge, or one that by incorporating the explanandum in the background knowledge it enhances its unity, offers a better understanding and, hence has more explanatory power.

The evaluation of explanatory power takes place in two directions. The *first* is to look at the specific background information (beliefs) which operate in a certain application of IBE. The *second* is to look at a number of structural features (standards) which competing explanations might possess. The prime characteristic of IBE is that it cannot operate in a 'conceptual vacuum', as Ben-Menahem (1990, 330) put it. Whatever else one thinks of an explanation, it must be such that it establishes some causal-nomological connection between the explanandum and the explanans. The details of this connection – and hence the explanatory story told – will be specified relative to the available background knowledge. To say that a certain hypothesis H is the best explanation of the evidence is to say, at least in part, that the causal-nomological story that H tells tallies best with background knowledge. This knowledge must contain all relevant information about, say, the types of causes that, typically, bring about certain effects, or the laws that govern certain phenomena etc. At least in non-revolutionary applications of IBE, the relevant background knowledge can have the resources to discriminate between better and worse potential explanations of the evidence. So, the explanatory power of a potential explanation depends on what other substantive information there is available in the background knowledge.[15] Let me call 'consilience' this feature of IBE which connects the background knowledge with the potential explanation of the evidence.

*Consilience*: Suppose there are two potentially explanatory hypotheses $H_1$ and $H_2$ but the relevant background knowledge favours $H_1$ over $H_2$. Unless there are specific reasons to challenge the background knowledge, $H_1$ should be accepted as the best explanation.

To a certain extent, there is room for a structural specification of the best explanation of a certain event (or piece of evidence). There are structural standards of explanatory merit which mark the explanatory power of a hypothesis and which, when applied to a certain situation, rank competing explanations in terms of their explanatory power. These standards operate crucially when the substantive information contained in the relevant background knowledge cannot forcefully discriminate between competing potential explanations of the evidence. The following list, far from being complete, is an indication of the relevant standards.[16]

*Completeness*: Suppose only one explanatory hypothesis H explains all data. That is, all other competing explanatory hypotheses fail to explain some of the data, although they are not refuted by them. H should be accepted as the best explanation.

*Importance*: Suppose two hypotheses $H_1$ and $H_2$ do not explain all relevant phenomena, but that $H_1$, unlike $H_2$, explains the most salient phenomena. Then $H_1$ is to be preferred as a better explanation.

*Parsimony*: Suppose two composite explanatory hypotheses $H_1$ and $H_2$ explain all data. Suppose also that $H_1$ uses fewer assumptions than $H_2$. In particular, suppose that the set of hypotheses that $H_1$ employs to explain the data is a proper subset of the hypotheses that $H_2$ employs. Then $H_1$ is to be preferred as a better explanation.

*Unification*: Suppose we have two composite explanatory hypotheses $H^k$ and $H^j$ a body of data $e_1, \ldots, e_n$. Suppose that for every piece of data

$e_i$ ($i = 1, \ldots, n$) to be explained $H^j$ introduces an explanatory assumption $H^j_i$ such that $H^j_i$ explains $e_i$. $H^k$, on the other hand, subsumes the explanation of all data under a few hypotheses, and hence it unifies the explananda. Then $H^k$ is a better explanation than $H^j$.

*Precision*: Suppose $H_1$ offers a more precise explanation of the phenomena than $H_2$, in particular an explanation that articulates some causal-nomological mechanism by means of which the phenomena are explained. Then $H_1$ is to be preferred as a better explanation.

Such standards have a lot of intuitive pull. Besides, they can characterise sufficiently well several instances of application of IBE in scientific practice (cf. Thagard 1978, 1988). But even if one granted that these standards have some genuine connection with explanatory quality or merit, one could question their epistemic status: why are they anything more than pragmatic virtues? If to call a certain virtue 'pragmatic' is to make it non-cognitive, to relegate it to a merely self-gratifying 'reason' for believing things, it should be clear that the foregoing explanatory virtues (standards) are not pragmatic. For they possess a straight cognitive function. As Paul Thagard (1989) has argued, such standards safeguard the explanatory coherence of our total belief corpus as well as the coherence between our belief corpus and a new potential explanation of the evidence. To say that a hypothesis that meets these standards has the most explanatory power among its competitors is to say that it has performed best in an explanatory coherence test among its competitors. Explanatory coherence is a cognitive virtue because, on some theories of justification at least, it is a prime way to confer justification on a belief or a corpus of beliefs (cf. BonJour 1985; Harman 1986). The warrant conferred on the chosen hypothesis, namely, that it fares better than others in an explanatory-quality test and that, as a result of this, it enhances the explanatory coherence of the belief corpus, is a defeasible warrant. But this is as it should be. The problem might be thought to be that there is no algorithmic way to connect all these criteria (with appropriate weights) so that they always engender a clear-cut ranking. The obvious rivalries among some of the criteria suggest that a lot of judgement should be exercised in this ranking. Such problems would be fatal only for those who thought that a suitable description of the method would have to be algorithmic, and in particular that it would have to employ a simple and universal algorithm. This aspiration should not have been taken seriously in the first place. Note also that although a simple and universal algorithm for IBE is not possible, there have been implementations of IBE, for example, by Thagard (1989) which employ a variety of algorithms. Besides, although IBE may be characterised at a very general and abstract level in the way presented above, there is good reason to think that many specific applications (e.g., in medical diagnosis) may employ important domain-specific criteria which require more careful empirical study.

### 10.3.2 Some philosophical issues

Some philosophers have expressed doubts about IBE which are based on the following worry: why should the information that a hypothesis is the best explanation of the evidence be a *prima facie reason* to believe that this hypothesis is true (or likely to be true)? As we have already seen in Chapter 6, Cartwright (1983, 4), has argued that the foregoing question cannot be successfully answered. Meeting this challenge will have to engage us in a proper understanding of the interplay between substantive background knowledge and considerations of explanatory coherence in rendering IBE a legitimate mode of inference.

What sort of *inference* is IBE? There are two broad answers to this. (1) We infer to the probable truth of the likeliest explanation insofar as and because it is the *likeliest* explanation. On this answer, what matters is how likely the explanatory hypothesis is. If it is likely we infer it; if it isn't we don't. (2) The best explanation, *qua* explanation, is likely to be true (or, at least more likely to be true than worse explanations). That is, the fact that a hypothesis H is the *best* explanation of the evidence issues a warrant that H is likely. Lipton (1991, 61–5), has noted that the first answer views IBE as an inference to the likeliest potential explanation, while the second views it as an inference to the loveliest potential explanation. The loveliest potential explanation is 'the one which would, if correct, be the most explanatory or provide the most understanding' (op. cit., 61). If we go for the likeliest potential explanation, Cartwright's challenge evaporates. For, best explanation and epistemic warrant are linked *externally* via some considerations of likelihood.[17] If there are reasons to believe that a certain hypothesis is likely (or the likeliest available), there is no further issue of epistemically warranted acceptance. But if we go for the likeliest potential explanation (i.e., the first answer above) then, IBE loses all of its excitement. For what is particularly challenging with IBE is the suggestion – encapsulated in answer (2) above – that the fact that a hypothesis is the *best* explanation (i.e., the loveliest one) *ipso facto* warrants the judgement that it is likely. If the loveliness of a potential explanation is shown to be a symptom of its truth, Cartwright's challenge is met in a significant and *internal* way.[18]

Lipton's own strategy has been to impose two sorts of filters on the choice of hypotheses. One selects a relatively small number of potential explanations as plausible, while the other selects the best among them as the actual explanation. Both filters should operate with explanatory considerations. Both filters should act as explanatory-quality tests. Still, although plausibility might have to do with explanatory considerations, why should plausibility have anything to do with likelihood? Lipton's answer is to highlight the *substantive assumptions* that need to be in place for IBE (as Inference to the loveliest potential explanation) to be possible. Explanatory considerations enter into the first filter (that of selecting a small number of hypotheses) by means of our substantive background knowledge that favours hypotheses that cohere well with (or are licensed by) our background beliefs (cf. Lipton 1991, 122). Insofar as these background beliefs are themselves likely, IBE operates within an environment of likely hypotheses. Given that the background beliefs themselves have been the product of past applications of IBE, they have been themselves imputed by explanatory considerations. So, the latter enter implicitly in the first filter and explicitly in the second (that of choosing the best among the competing hypotheses that are licensed by the background beliefs).

We can see the crux of all this by looking at Josephson's aforementioned schema (A) for IBE. The crucial judgement for the inference to take place is that no other hypothesis explains *D* as well as *H*. This judgement is the product of (a) filtering the competing hypotheses according to substantive background knowledge and (b) choosing among *them* by explanatory considerations. The upshot of all this is that the application of IBE relies on substantive background knowledge. Without it, IBE as an *inference* is simply impotent.[19] But notice that the structural features that render an explanation better than another are part and parcel of the background knowledge. They are just the more abstract part of it that tells us how to evaluate potential explanations. Notice also that these general structural features are complemented by particular ones when it comes to specific applications of IBE. As Josephson (1994, 14) has noted, in specific cases the likelihood of the chosen 'best explanation' H will depend on considerations such as 'how decisively H surpasses the alternatives' and 'how much confidence there is that all plausible explanations have been considered (how thorough was the search for alternative explanations)'.

Suppose that all this is not convincing. Suppose that we haven't made a case for the claim that the best (loveliest) explanation and the likeliest explanation may reasonably be taken to coincide in light of the relevant background knowledge. There is still an indirect answer available to Cartwright's challenge. Note that we are concerned with the *prima facie* warrant for accepting a hypothesis H. The question then is: is the fact that H is rendered the best explanation of the evidence a *prima facie* reason for its acceptance? If, following Pollock (1986, 124), we view justification as 'epistemic permissibility', it is obvious that the answer to the foregoing question can only be positive. For to say that the fact that H is the best explanation of the evidence is a *reason* for the acceptance of H is to say that (a) it is all right (i.e., it is permissible) to believe in H on this basis; and (b) that this permissibility is grounded on the explanatory connection between H and the evidence. It is this explanatory connection which makes the acceptance of H *prima facie* reasonable since it enhances the coherence of our total belief corpus. By incorporating H into our belief corpus BC as the best explanation of the evidence we enhance the capacity of BC to deal with new information and we improve our understanding not just of why the evidence is the way it is but also of how this evidence gets embedded in our belief corpus.

To see how all this works out, note the following. It is explanatory (causal-nomological) connections which hold our belief corpus together. It is such connections which organise the individual beliefs that form it and make the corpus useful in understanding, planning, anticipating etc. (cf. Harman 1986). Faced with a choice among competing explanatory hypotheses of some event, we should appeal to reasons to eliminate some of them.[20] Subjecting these hypotheses to an explanatory-quality test is the prime way to afford these reasons. Those hypotheses that fare badly in this test get eliminated. By having done badly in the test, they have failed at least some of the intuitively compelling criteria of explanatory power. They have either failed to cohere well with the relevant background information, or have left some of the data unaccounted for, or have introduced gratuitous assumptions into the explanatory story, or what have you. If this test has a clear winner (the best explanation), this is the only live option for acceptance. In the end, IBE enhances the explanatory coherence of a background corpus of belief by choosing a hypothesis which brings certain pieces of evidence into line with this corpus. And it is obviously reasonable to do this enhancement by means of the best available hypotheses. This coherence-enhancing role of IBE, which has been repeatedly stressed by Harman (1979, 1986, 1995), William Lycan (1988) and Thagard (1978, 1989), is ultimately the warrant-conferring element of IBE.

Some philosophers think that there may be a tension between the two prime aspects of IBE that I have described above, namely, its reliance on considerations of explanatory coherence and its dependence on substantive background beliefs. Timothy Day and Harok Kincaid (1994, 275) for instance, argue that if IBE is primarily seen as relying on considerations of explanatory coherence, it becomes 'redundant and uninformative'. It reduces to 'nothing more than a general admonition to increase coherence (ibid., 279). And if IBE is primarily seen as being dependent on substantive background knowledge, it 'does not name a fundamental pattern of inference' (ibid., 282). Rather, they argue, it is an instance of a strategy 'that infers to warranted beliefs from background information and the data', without necessarily favouring an explanatory connection between hypotheses and the data (cf. ibid.). Day and Kincaid favour a *contextual* understanding of IBE, since, they say, it has 'no automatic warrant' and its importance 'might well differ from one epistemic situation to the next' (ibid., 282).

I think that (a) the two aspects of IBE are not in any tension; and (b) they engender a rather general and exciting mode of ampliative reasoning. Certainly, more work needs to be done on the notion of coherence and its link with explanation. But if we adopt what Lycan (1989) has called 'explanationism', it should be clear that explanatory coherence is a vehicle through which an inference is performed and justified. IBE is the mode of inference which achieves ampliation via explanation and which licenses conclusions on the basis of considerations which increase explanatory coherence. Yet, it is wrong to think that the achievement (or enhancement) of explanatory coherence is just a formal-structural matter. Whatever else it is, the best explanation of the evidence (viz., the one that is the best candidate for an enhancement of the explanatory coherence of a belief corpus) has some substantive content which is constrained (if not directly licensed) by the relevant substantive background knowledge. So, substantive background information is not just the material on which some abstract considerations of explanatory coherence should be imposed. It is also the means by which this coherence is achieved. To infer to the best explanation H of the evidence is to search within the relevant background knowledge for explanatory hypotheses and to select the one (if there is one) which makes the incorporation of the evidence into this background corpus the most explanatorily coherent one. The selection, as we have seen, will be guided by both the substantive background knowledge and some relatively abstract structural standards. That this process is not an inference can be upheld only if one entertains the implausible views that to infer is to deduce and that to infer is to have 'an automatic warrant' for the inference. Not all changes in the background knowledge will be based on explanatory considerations. But given that some (perhaps most) are, IBE will have a distinctive (and exciting) role to play.

To sum up: the *prima facie* reasonableness of IBE cannot be seriously contested. Even if one can question the link between best explanation and truth, one cannot seriously question that the fact that a hypothesis stands out as the best explanation of the evidence offers defeasible reasons to warrantedly accept this hypothesis.[21]

### 10.3.3 Abduction and the two desiderata

This preliminary defence of the reasonableness of IBE was necessary in order to dispel some natural doubts towards it. Presently, we need to see how IBE fares vis-à-vis EI and HD. I will suggest that both EI and HD are extreme cases of IBE, but while EI is an interesting limiting case, HD is a degenerate one whose very possibility shows why IBE is immensely more efficient. Besides, I will argue that IBE has all the strengths and none of the weaknesses of either EI or HD.

That proper inductive arguments are instances of IBE has been argued by Harman (1979) and been defended by Josephson (1994, 2000) and Psillos (2000a). The basic idea is that good inductive reasoning involves comparison of alternative potentially explanatory hypotheses. In a typical case, where the reasoning starts from the premise that 'All As in the sample are B', there are (at least) two possible ways in which the reasoning can go. The first is to withhold drawing the conclusion that 'All As are B', even if the relevant predicates are projectable, based on the claim that the observed correlation in the sample is due to the fact that the sample is biased. The second is to draw the conclusion that 'All As are B' based on the claim that the observed correlation is due to the fact that there is a nomological connection between being A and being B such that All As are B. This second way to reason implies (and is supported by) the claim that the observed sample is not biased. What is important in any case is that which way the reasoning should go depends on explanatory considerations. Insofar as the conclusion 'All As are B' is accepted, it is accepted on the basis that it offers a better explanation of the observed frequencies of As which are B in the sample, in contrast to the (alternative potential) explanation that someone (or something) has biased the sample. Insofar as the generalisation to the whole population is not accepted, this judgement will be based on providing reasons that the biased-sample hypothesis offers a better explanation of the observed correlations in the sample. Differently put, EI is a limiting case of IBE in that (a) the best explanation has the form of a nomological generalisation of the data in the sample to the whole relevant population and (b) the nomological generalisation is accepted, if at all, on the basis that it offers the best explanation of the observed correlations on the sample. HD, on the other hand, is a limiting but degenerate case of IBE in the following sense: if the only constraint on an explanatory hypothesis is that it deductively entails the data, any hypothesis which does that is a potential explanation of the data. If there is only one such hypothesis, it is automatically the 'best' explanation. But it is trivially so. The very need for IBE is suggested by the fact that HD is impotent, as it stands, to discriminate between competing hypotheses which entail (and hence explain in this minimal sense) the evidence.

How, then, does IBE fare vis-à-vis the two desiderata for the method, namely, ampliation and epistemic warrant? Remember that EI is minimally ampliative and maximally epistemically probative, whereas HD is maximally ampliative and minimally epistemically probative. Like HD, IBE is maximally ampliative: it allows for the acceptance of hypotheses that go far beyond the data not just in a horizontal way but also in a vertical one. And given that EI is a special case of IBE, IBE can – under certain circumstances – be as epistemically probative as EI. But unlike HD, IBE can be epistemically probative in circumstances that HD becomes epistemically too permissive. For IBE has the resources to deal with the so-called 'multiple explanations' problem (cf. Psillos 2000a, 65). That is, IBE can rank competing hypotheses which all, *prima facie*, explain the evidence in terms of their explanatory power and therefore evaluate them. In order to see how this evaluative dimension of IBE can issue in epistemic warrant, let us examine the types of defeaters to the reasons offered by IBE.

Recall from Section 10.2.1 that to say that one is *prima facie* warranted to accept the outcome of an ampliative method is to say that one has considered several possible defeaters of the reasons offered for this outcome and has shown that they are not present. If this is done, there are no *specific* doubts about the warrant for the outcome of the method. Recall also that there are two general types of defeater, rebutting and undercutting ones. If there is an observation which refutes the best explanation of the evidence so far, this is a rebutting defeater of the best explanation. But IBE fares better than HD vis-à-vis the Duhem–Quine problem. For, although any hypothesis can be saved from refutation by suitable adjustments to some auxiliary assumptions (and hence although any rebutting defeater can be neutralised), IBE can offer means to evaluate the impact of a recalcitrant piece of evidence on the conclusion that the chosen hypotheses is the best explanation of the evidence. HD does not have the resources to perform this evaluation. If the sole constraint on the acceptance of the hypothesis is whether or not it entails the evidence, it is clear that a negative observation can only refute the hypothesis. If the hypothesis is to be saved, the blame should be put on some auxiliaries, but – staying within HD – there is no independent reason to do so. In IBE, the required independent reasons are provided by the relevant explanatory considerations: if there are strong reasons to believe that a hypothesis is the best explanation of the evidence, there is also reason to stick to this hypothesis and make the negative observation issue in some changes to the auxiliary assumptions. After all, if a hypothesis has been chosen as the best explanation, it has fared best in an explanatory-quality test with its competing rivals. So unless there is reason to think that it is superseded by an even better explanation, or unless there is reason to believe that the recalcitrant evidence points to one of the rivals as a better explanation, to stick with the best explanatory hypothesis is entirely reasonable.

This last thought brings us to the role of undercutting defeaters in IBE. Recall that in the case of HD, any other hypothesis which entails the same evidence as H is an undercutting defeater for (the warrant for) H. Given that there are going to be a lot of such alternative hypotheses, the warrant for H gets minimised. In IBE it is simply not the case that any other hypothesis which entails the evidence offers an explanation of it. It is not required that the explanatory relation between the evidence and the hypothesis be deductive.[22] Even if we focus on the special case in which this relation is deductive, IBE dictates that we should look beyond the content of each potential explanatory hypothesis and beyond the relations of deductive entailment between it and the evidence in order to appraise its explanatory power. Two or more hypotheses may entail the same evidence, but one of them may be a better explanation of it. So, the presence of a worse explanation cannot act as an undercutting defeater for the acceptance of the best explanatory hypothesis. The choice of the best explanation has already involved the consideration of possible undercutting defeaters (viz., other potential explanations of the evidence) and has found them wanting. The judgement that a certain hypothesis is the best explanation of the evidence is warranted precisely because it has rested on the examination and neutralisation of possible undercutting defeaters.

To be sure, IBE is defeasible. The discovery of a better explanation of the evidence will act as an undercutting (sometimes even as a rebutting defeater) of the chosen hypothesis. But this is harmless for two reasons. First, given the information available at a time *t*, it is reasonable to infer to the best available explanation H of the present evidence even if there may be even better possible explanations of it. The existence of hitherto unthought of explanations is a contingent matter. H has fared in the explanatory-quality test better than its extant competitors. Hence it has neutralised a number of possible undercutting defeaters. That there may be more possible undercutting defeaters neither can it be predicted, nor can it retract from the fact that it is *prima facie* reasonable to accept H. If the search for other potential explanations has been thorough, and if the present information does not justify a further exploration of the logical space of potentially explanatory hypotheses, there is no *specific* reason to doubt that the current best explanation is simply the best explanation. If such doubts arise later on they are welcome, but do not invalidate our present judgement.

The natural conclusion of all this is that IBE admits of clear-cut undercutting defeaters, but unlike HD it has the resources to show when a potential undercutting defeater can be neutralised. And it also admits of clear-cut rebutting defeaters, but unlike HD it can explain how and why such a possible defeater can be neutralised. So, when it comes to its epistemically probative character, IBE can reach the maximal epistemic warrant of EI (since EI is an extreme case of IBE), but it goes far beyond the minimal epistemic warrant of HD (since it offers reasons to evaluate competing hypotheses in an explanatory-quality test). When it comes to ampliation, like HD and unlike EI, it reaches up to maximal ampliation (cf. the following chart).

| | EI | HD | IBE |
|---|---|---|---|
| Ampliation | Minimal | Maximal | Maximal |
| Epistemic Warrant | Maximal | Minimal | Far more than minimal and up to maximal |

Abduction, understood as inference to the best explanation, satisfies in the best way the two desiderata of ampliation and epistemic warrant and also strikes the best balance between the role that background knowledge plays in ampliative reasoning and the role that explanatory considerations (as linked with the demand of explanatory coherence) play in justifying an inference. Obviously, more work needs to be done on the notion of explanatory coherence and also on the role of coherence in justification. The good news, however, is that IBE can emerge as the general specification of scientific method which promises to solve in the best way its central philosophical problem.

# 11
# Inference to the Best Explanation and Bayesianism

Lately, there has been a lot of discussion about the place of IBE in Bayesian reasoning. Niiniluoto (2004, 68) argues that 'Bayesianism provides a framework for studying abduction and induction as forms of ampliative reasoning'. There is a tension, however, at the outset. Bayesian reasoning does *not* have rules of acceptance. On a strict Bayesian approach,[1] we can never detach the probability of the conclusion of a probabilistic argument, no matter how high this probability might be. So, strictly speaking, we are never licensed to accept a hypothesis on the basis of the evidence. All we are entitled to do, we are told by strict Bayesians, is (a) to detach a conclusion about a probability, namely, to assert that the posterior probability of a hypothesis is thus and so; and (b) to keep updating the posterior probability of a hypothesis, following Bayesian conditionalisation on fresh evidence. But IBE is a rule of acceptance. In its least controversial form, IBE authorises the *acceptance* of a hypothesis H, on the basis that it is the best explanation of the evidence. Think of the standard IBE-based argument for the existence of middle-sized material objects. According to this, the best explanation of the systematic, orderly and coherent way we experience the world is that there are stable middle-sized material objects that cause our experiences. Presumably, those who endorse this argument do not just assert a conclusion about a probability; they assert a conclusion, *simpliciter*. Their claim is not that the probability that material objects exist is high, but rather that it is reasonable to accept that they do exist. Hence, there is a tension between Bayesianism and standard renderings of IBE. This might make us wary of attempts to cast IBE in a Bayesian framework. But this is only the beginning of our worries.

Niiniluoto (2004) surveys a variety of recent results about the connection between IBE and Bayesian confirmation. They are all invariably instructive. But I want to challenge the *motivation* for attempting this. There are two questions to be asked. *First*, if we were to cast IBE within a Bayesian framework, could we do it? I do not doubt that this can be done (given the flexibility of the Bayesian framework). But as I shall raise some worries about the ways that can done. These worries will usher in the need the raise a *second* question, namely, should we want to cast IBE within a Bayesian framework? This question, I think, is more intriguing than the first.

## 11.1 IBE and Bayesian kinematics

The crux of IBE, no matter how it is formulated, is that explanatory considerations should inform (perhaps, determine) what is reasonable to believe. There are several ways to import explanatory considerations into a Bayesian scheme. There is a contentious one, due to van Fraassen (1989). He claims that the right way to cast IBE within a Bayesian framework is to give *bonuses* to the posterior probabilities of hypotheses that are accepted as best explanations of the evidence. That is, *after* having fixed the posterior probability of a hypothesis in a Bayesian way, if this hypothesis is seen as the best explanation of the evidence, its posterior probability is risen. It's not hard to see that if one followed this way of updating one's degrees of belief, one would end up with a Dutch book. In fact, this is precisely the strategy that van Fraassen himself follows in order to argue that, seen as a rule of updating degrees of belief, IBE is incoherent. But why should one take van Fraassen's suggestion seriously? As noted already, the key recommendation of IBE is that explanatory considerations should inform (perhaps, determine) what we reasonably come to believe. If one were to cast IBE within a Bayesian framework, one should make sure that explanatory considerations are *part* of the Bayesian kinematics for the determination of the posterior probability of a theory, not something that should be *added* on to confer bonus degrees of belief to the end product.

Given the Bayesian machinery, there are two ways in which explanatory considerations can be *part* of the Bayesian kinematics. They should either be reflected in the prior probability of a theory, relative to background knowledge, or (inclusively) in the likelihood of the theory. Though priors and likelihoods *can* reflect explanatory judgements, it is clear that they fail to discriminate among competing hypotheses with the *same* priors and the *same* likelihoods. This problem is particularly acute for the case of likelihoods. The whole point of insisting on IBE is that it promises rationally to resolve observational ties. When two or more competing hypotheses entail the very same evidence, their likelihoods will be the same. Hence, likelihoods cannot resolve, at least some (perhaps the most significant), observational ties.

Actually, things get worse if we base our hopes on likelihoods. One thought, explored by Niiniluoto (2004, 76–7) might be to equate the best explanation with the hypothesis that enjoys the highest likelihood. But, as we are about to see, this is deeply problematic. In fact, as the base-rate fallacy shows, likelihoods are relatively mute. If explanatory considerations enter the Bayesian story via likelihoods, so much the worse for the explanatory considerations.

## 11.2 Likelihoods and the base-rate fallacy

We have already discussed the base-rate fallacy in Chapter 3, Section 3.3. It shows that it is incorrect just to equate the best explanation of the evidence with the hypothesis that has the highest likelihood. It turns out that, if we consider solely the likelihood of a hypothesis, and if we think that this is the way to determine the best explanation, there is no determinate answer to the question 'what is the best explanation of the evidence?' A very small prior probability can dominate over high likelihood and lead to a very small posterior probability. Let me put the point in a more conspicuous way. If we try to cast IBE within a Bayesian framework by focusing on likelihoods (i.e., by saying that the best explanation is the hypothesis with the highest likelihood), intuitive judgements of best explanation and judgements of Bayesian confirmation may well come apart.

The point about likelihoods generalises. Consider what is called the Bayes factor, that is, the ratio of likelihoods prob(e/not–H)/prob(e/H) – see Chapter 3, Section 3.4. One might try to connect IBE with likelihoods as follows. If the Bayes factor is small, H is a better explanation of the evidence *e* than not–H. For, the thought will be, there are two ways in which the Bayes factor can be minimised: either when *e* is very unlikely when H is false or when *e* is very likely when H is true. Here is a way to write Bayes's theorem:

$$\text{prob}(H/e) = \text{prob}(H)/\text{prob}(H) + f\ \text{prob}(\text{not–}H),$$

where $f$ is the Bayes factor, that is, prob(e/not–H)/prob(e/H). Wouldn't we expect that the smaller the Bayes factor is, the greater is the posterior probability of the hypothesis? Wouldn't we thereby find a way to accommodate IBE, via the Bayes factor, within Bayesianism? As explained in

Chapter 3, Section 3.4, what really happens depends on the prior probability. On its own, the Bayes factor tells almost nothing. I say *almost* nothing because there is a case in which the prior probability of a hypothesis does not matter. This is when the Bayes factor is *zero*. Then, no matter what the prior prob(H) is, the posterior probability prob(H/e) is one. So, it's only when just one theory can explain the evidence (in the sense that the likelihood prob(e/not–H) is zero) that we can dispense with the priors. That's a significant result. But does it show that IBE is accommodated within Bayesianism? In a sense, it does, and yet this sense is not terribly exciting. If there was only one potential explanation, it would be folly not to accept it. This case, however, is really exceptional. We are still left with the need to distinguish between grue and green!

The moral so far is double. On the one hand, likelihoods cannot capture the notion of a good (the best) explanation. Differently put, even if likelihoods could, to some extent, carry the weight of explanation, they couldn't carry *all* of this weight on their own. On the other hand, we need to take into account the prior probabilities before we draw safe conclusions about the degree of confirmation of a hypothesis.

## 11.3 Explanation and prior probabilities

What then remains of the Bayesian kinematics as an (indispensable) entry point for explanatory judgements is the prior probabilities. It is one thing to say that priors are informed by explanatory considerations and quite another thing to say that they *should* be so informed. No-one would doubt the former, but subjective Bayesianism is bound to deny the latter. So, we come to the crux. There are two ways to think of IBE within a Bayesian framework. The first pays only lip service to explanatory considerations. For all the work in degree-of-belief updating (or, as some Bayesians say, in maintaining internal coherence in an agent's belief-corpus) is done by the usual Bayesian techniques and, perhaps, by the much-adored appeal to the washing out of priors. It may be admitted that the original assignment of prior probabilities might be influenced by explanatory considerations but the latter are no less idiosyncratic (from the point of view of the subjective Bayesian) than specifying the priors by, say, consulting a soothsayer. If we think this way, IBE is rendered consistent with Bayesianism, but it loses much of its excitement. It just amounts to a *permission* to use explanatory considerations in fixing an initial distribution of prior probabilities.

The other way to think of IBE within a Bayesian framework is to take explanatory considerations to be a *normative* constraint on the specification of priors. This is the way I would favour, if I were to endorse the Bayesianisation of IBE. It would capture the idea that explanatory considerations should be a rational constraint on inference. We might still be short of acceptance, since all we end up with is a degree of belief (no matter how high), but it would, at least, be a degree of *rational* belief. This move would also show how the resolution of observational ties is not an idiosyncratic matter. For some theories would command a higher initial rational degree of belief than others and this would be reflected, via Bayesian kinematics, in their posterior probability.[2]

But don't we all know that the story I have sketched is, to say the least, *extremely* contentious? It would call for an objectivisation of subjective Bayesianism. Whence do the explanatory virtues get their supposed rational force? And how are they connected with truth? These are serious worries. I am not sure they are compelling. For instance, I think that there can be an a posteriori argument to the effect that theories with the explanatory virtues are more likely to be true than others (cf. Psillos 1999, 171–6). And there is also an argument, discussed in Chapter 10, to the effect that judgements of prior probabilities should aim to improve the coherence of our system of beliefs and that the explanatory virtues improve such coherence. But showing all this would be an uphill battle. It would cal for a radical departure from the standard Bayesian criteria of rationality and belief-revision. The project of accommodating IBE within Bayesianism would involve a radical rethinking of Bayesianism. And not many people are, nowadays, willing for such a radical rethinking.

## 11.4 A dilemma

The way I have described things leads us to a dilemma. *Either* accommodate (relatively easily) IBE within Bayesianism but lose the excitement and most of the putative force of IBE *or* accommodate an interesting version of IBE but radically modify Bayesianism. I guess we all agree that Bayesianism is the best theory of confirmation available. But at least some of us are unwilling to think that Bayesianism is the final word on the matter, since we think that there is more to rationality (and to scientific method) than Bayesians allow.[3] Those of us who are friends of IBE might then have to reject the foregoing dilemma altogether. This would bring us back to the second question I raised in the beginning of the chapter: *should we want to cast IBE within a Bayesian framework?*

I cannot start answering this question here. I hope to have sketched why there are reasons to take it seriously. I will conclude with a note on how a *negative* answer to it can be motivated. IBE is supposed to be an *ampliative* method of reasoning. It is supposed to deliver informative hypotheses and theories, namely, hypotheses and theories which exceed in content the observations, data, experimental results etc. which prompt them. This content-increasing aspect of IBE is indispensable, if science is seen, at least *prima facie*, as an activity that purports to extend our knowledge (and our understanding) beyond what is observed by means of the senses. Now, Bayesian reasoning is *not* ampliative. In fact, it does not have the resources to be ampliative. All is concerned with is maintaining synchronic consistency in a belief corpus and (for some Bayesians, at least) achieving diachronic consistency too. Some Bayesians, for example, Howson (2000), take probabilistic reasoning to be a mere extension of deductive reasoning, which does not beget any new factual content.

There might be two related objections to what I have just said. The first might be that Bayesian reasoning allows for ampliation, since this can be expressed in the choice of hypotheses over which prior probabilities are distributed. In other words, one can assign prior probabilities to ampliative hypotheses and then use Bayesian kinematics to specify their posterior probabilities. The second (related) objection may be found in what Niiniluoto says at some point, namely, that 'the Bayesian model of inference helps to show how evidence may confirm hypotheses that are abductively introduced to explain them' (Niiniluoto 2004, 71). Here again, the idea is that abduction might suggest ampliative hypotheses, which are then confirmed in a Bayesian fashion. If we elaborate and combine the two objections in an obvious way, they imply the following: ampliative IBE and non-ampliative Bayesian reasoning might well work in tandem to specify the degree of confirmation of ampliative hypotheses.[4]

I too have toyed with this idea and still think that there is something to it. In an earlier piece I noted:

> although a hypothesis might be reasonably accepted as the most plausible hypothesis based on explanatory considerations (abduction), the *degree of confidence* in this hypothesis is tied to its degree of subsequent confirmation. The latter has an antecedent input, i.e., it depends on how good the hypothesis is (i.e., how thorough the

> search for other potential explanations was, how plausible a potential explanation is the one at hand etc.), but it also crucially depends on how well-confirmed the hypothesis becomes in light of further evidence. So, abduction can return likely hypotheses, but only insofar as it is seen as an integral part of the method of inquiry, whereby hypotheses are further evaluated and tested. (Psillos 2000a, 67)

But we can also see the limitations of this idea. What, in effect, is being conceded is that IBE (or abduction) operates *only* in the context of discovery, as a means to generate plausible ampliative hypotheses and to distil the best among them. Then, the best explanation is taken over to the context of justification, by being embedded in a framework of Bayesian confirmation, which determines its credibility. I think the friends of IBE have taken IBE to be both ampliative *and* warrant-conferring at the same time. It is supposed to be an ampliative method that confers warrant on the best explanation of the evidence. If we are concerned with giving a precise degree of confirmation to the best explanation, or if we subject it to further testing, we can indeed embed it in a Bayesian framework. But something would have gone amiss if we thought that the best explanation was *not* reasonably acceptable before it was subjected to Bayesian confirmation. How this reasonable acceptance could be analysed has already been explained in the previous chapter.

# Notes

## Introduction

1. Here my own reading of Kant is influenced by Rae Langton's (1998). I do not however get into interpretative issues concerning the character of this dichotomy, and in particular whether it implies a two-world or a two-aspect approach to the things-in-themselves (see Allison 2004).
2. For a set of critical perspectives on Psillos (1999), see the book symposium 'Quests of a Realist', *Metascience* **10** (2001), pp. 341–70 (co-symposiasts: Michael Redhead; Peter Lipton, Igor Douven & Otavio Bueno) and my reply to them – Psillos (2001).

## 1 The present state of the scientific realism debate

1. 'Theoretical truth' is an abbreviation for 'truth of theoretical assertions', namely, those assertions that are couched in a theoretical vocabulary. Similarly for 'observational truth'.
2. Its problems have been shown by Miller (1974) and Tichy (1974).
3. It's a bit ironic that van Fraassen (1980, 8) also characterises realism as an axiological thesis together with a (non-Popperian) doxastic attitude, namely, that acceptance of a theory involves belief in its truth.
4. For a fresh look into Putnam's (1981) 'internal realism' cf. Niiniluoto (1999, Chapter 7).
5. Quine (cf. 1981, 92), however, is a professed realist. For an exploration of Quine's views on realism see Hylton (1994).
6. Jennings (1989) has tried to articulate 'scientific quasi-realism' in the spirit of Blackburn (1984). The main thought is that quasi-realists can 'earn the right' to talk about the truth-or-falsity of theories, without the concomitant commitments to a realist ontology: the posited theoretical entities inhabit a 'projected' world. That this is a genuine middle way has been challenged by Fine (1986b) and Musgrave (1996).
7. Two reviewers of Leplin's book (Sarkar 1998; Ladyman 1999) argue that the Uniqueness Condition (UC) is too strong: given that some other theory T′ may predict O soon after T has first predicted it, why should we accept that the (accidental) historical precedence of T makes O novel for T but not for T′? In fairness to Leplin, it is crucial that T′ satisfies the Independence Condition (IC). If not, T′ is not a contender. If yes, O can be said to be *prima facie* novel for both T and T′. Then, naturally, the failure of UC cannot make O shift the epistemic balance in favour of either T or T′ and more evidence should be sought after.
8. Van Fraassen challenges the realists' epistemic optimism. But, unlike Wright and like scientific realists, he takes scientific statements 'to have

truth-conditions entirely independently of human activity or knowledge' (van Fraassen 1980, 38).

9. For a critique of some social constructivists' reaction to UTE, cf. Laudan (1996, 50–3).
10. Van Fraassen (1989, 160–70) suggested that IBE – conceived as a rule – is incoherent. Harman (1996) and Douven (1999) have rebutted this claim. See also Psillos (2007b).
11. Apart from noting that persistent retention at the theoretical level may be a reliable (but fallible) way to single out the theoretical assertions that won't suffer truth-value revision, one can claim that Almeder's stance falls pray to the 'preface paradox'. Almeder (1992, 180) tries to counter this claim. But a related problem still remains. A 'blind realist' asserts *both* that for each and every theoretical assertion P we can't know whether it is true *and* that we know that some Ps are true. Even if this joint assertion is consistent, the first part removes any basis for accepting the second.
12. And, of course, Fine's (1986a, 1986b) quietist dismissal of the philosophical debate altogether. Fine's views are criticised in detail in Musgrave (1989) and my work (Psillos 1999, Chapter 10).
13. Van Fraassen (1985, 255) implies that since the probability of a theory's being empirically adequate is less than or equal to its probability of being true, belief in truth is 'supererogatory'. But the above probabilistic relation between truth and empirical adequacy doesn't imply that the probability that the theory is true is not (or cannot be) high enough to warrant belief. What realists emphatically deny is that theoretical assertions are inherently insupportable. A variant of this thought is explored in Dorling (1992, 367–8).
14. This has been rectified since these words were first written. The volume by Bradley Monton (2007) contains a number of papers with a thorough discussion of van Fraassen's new epistemology, with replies by van Fraassen. See also Psillos (2007b).
15. A shorter version of the review of Chakravartty's book (which was co-authored with Dimitris Papayannakos) will appear in *Isis*.
16. The review of Turner's book appeared in *Notre Dame Reviews of Books* (2008.05.11).
17. A version of this review appeared in *Journal of Critical Realism*, 4 (2005), 255–61.

## 2 Scientific realism and metaphysics

1. Given the close link between propositions and facts, it's important to understand facts as the truth-makers of true propositions.
2. There is another way to be an antirealist. This is to say that the propositions of the contested class are not *really* propositions: they are not apt for truth and falsity; they cannot admit of truth-values. Traditional syntactic instrumentalism, ethical non-cognitivism and mathematical formalism might be classified under this view. I would call this view non-factualism (since the contested propositions are said not to be in the business of describing facts) and would distinguish it from anti-factualism (which says that the contested propositions are *false*). But I will not discuss non-factualism further. For an important

attempt to describe and challenge the metaphysics of non-factualism, see Devitt (2001).

3. Devitt (2001, 591–2) has recently come very close to accepting a role for a substantive notion of truth in the characterisation of realism. It concerns what he calls 'atypical realism', the view that there are facts that make a set of propositions true but that these facts are not further explainable (or grounded). I claim that the general characterisation of realism must be broad enough to allow 'atypical' realists to be realists without any guilt.
4. In Psillos (2006a) I claim that current science does not commit us to the view that the properties of the fundamental particles are ungrounded powers.

## 3 Thinking about the ultimate argument for realism

1. I have tried to explore some of these problems in Psillos (2007b).
2. A recent paper which casts fresh light on the role of the NMA in the realism debate is Ghins (2001).
3. This, in effect, sums up premises (ii) to (iv) of (*IBE*).
4. For more on non-deductive reasoning and on the way IBE should be understood as a *genus* of ampliative reasoning, see Chapter 10 and Psillos (2007a).
5. Musgrave might reply to this by saying that scientists employ 'demonstrative inductions', which are really deductions, though not deductions *from* the phenomena, as Newton (cf. 1999a, 303 & 306) thought. I don't want to discuss this issue here, though it certainly needs attention. Briefly put, the thrust of demonstrative induction is that premises of greater generality and premises of lesser generality will yield a conclusion of intermediate generality. But this must be noted: it is wrong to think that demonstrative induction frees us from the need to engage in ampliative inference. As Norton (1994, 12) notes: 'Typically, ampliative inference will be needed to justify "the premises of greater generality".'
6. This formulation does not exactly match the way Howson puts the argument, but closely resembles it.
7. By Bayes's theorem, prob(H/+) = prob(+/H)prob(H)/prob(+), where prob (+) = prob(+/H)prob(H)+prob(+/−H)prob(−H). Plug in the following values: prob(+/H) = 1, prob(H) = .001, prob(−H) = .999, prob(+/−H) = .05. Then, prob(H/+) is roughly equal to .02.
8. This problem was first investigated by Tversky and Kahneman (1982). It was dubbed 'the base-rate fallacy' by Bar-Hillel (1980).
9. To be more precise, we need to state the conclusion thus: Therefore, $\text{prob}_{\text{new}}(T)$ is high, where $\text{prob}_{\text{new}}(T) = \text{prob}_{\text{old}}(T/S)$.
10. In connection with the base-rate fallacy, L.J. Cohen (1981) has made the general point that there is no such thing as *the* relevant base-rate.
11. There is a worry here, voiced by Levin (1984), namely, that the truth of the theory does not explain its success. He asks: '(w)hat kind of *mechanism* is truth? How does the truth of a theory bring about, cause or create, its issuance of successful predictions? Here, I think, we are stumped. Truth (...) has nothing to do with it' (ibid., 126). Musgrave (1999a, 68–9) has answered

this worry effectively. What does the explaining is the theory. But, Musgrave adds: 'Semantic *ascent* being what it is, we do not have *rival* explanations here, but rather equivalent formulations of the *same* explanation. '*H* believed that *G* and *G*' is equivalent to '*H* believed *truly* that *G*' (given the theory of truth that Levin and the realists both accept' (ibid., 69). He then goes on to claim, correctly I think, that the explanation of the success of an action in terms of the *truth* of the agent's relevant beliefs is a mechanical or causal explanation.

12. If probability .59 is too low to capture the court's call that the case should be proven 'beyond reasonable doubt', we can alter the numbers a bit so that the probability that the cab was green is high enough.
13. A similar point is made by Windschitl and Wells (1996, 41).
14. The base-rate fallacy has been subjected to very detailed and informative scrutiny by Jonathan Koehler (1996).
15. I don't want to deny that high probability is sufficient for warranted belief. But is it necessary? I don't think so. One of the prime messages of the statistical relevance model of explanation is that *increase* in probability does count for warranted belief. Now, empirical success does increase the probability of a theory's being approximately true, even with a low base-rate for truth. This can be easily seen by looking again at the example that preceded argument ($B^1$). There, the prior probability prob(T) of T was 1 per cent but the posterior probability prob(T/S) rose to 17 per cent. So, success does make a difference to the probability of theory's being true.
16. For a critique of likelihoodism, see Achinstein (2001, 125–31).

## 4 Against neo-instrumentalism

1. Stanford (2006, 167–8) ponders a somewhat similar line of thought on behalf of the realist, takes it to be promising, but dismisses it on the grounds that it is unconvincing: it is merely one potential explanation among others, including Stanford's own, namely, that our judgements about the truth of past theories and our judgements about their successes have a *common source*. I fail to feel the weight of this point. Stanford's own potential explanation is external: it tells us something about the source of the scientists' (or of the realists') judgements, namely, that this source is current theory. Even if true, this line is compatible within an internal potential explanation of the emergence of a stable network of theoretical assertions along the realist lines, namely, along the lines that being part of this stable network is best explained by being truthlike.

## 5 Tracking the real: Through thick and thin

1. In Azzouni (2004), he avoids the expression 'thin epistemic access' to the real. However, the terminology is useful and I have decided it to keep it.
2. Though true, this is overstated. Science can proceed without observations at all, if by 'observation' we mean something akin to having a (conscious) observer perceiving (or experiencing) the data. As Feyerabend (1969) noted

long ago, science without experience *is* possible. For more on this, see Fodor (1991).

3. The same argument applies to Azzouni's (2004, 13) claim that 'background perceptual regularities' have been 'proved immune to epistemic corrosion'. This is certainly correct. But it does *not* follow from this that they get their epistemic authority independently of theory.
4. The sense of 'is' that Sellars has in mind here is explained in Sellars (1956, 314ff). It is the 'is' of theoretical identification.
5. This point is also made vividly by Hempel (1988).
6. A detailed defence of this point is given by Churchland (1979).
7. This may be misleading. As Sellars explains, these objects do not exist *as they are conceived* by the observational framework.
8. I have tried to offer an account of this theory in Psillos (1999, Chapter 8). I analyse and defend inference to the best explanation in Chapter 10.
9. For more on the differences between Kitcher and myself, see Psillos (1999, 111–12).
10. Azzouni (2004, 386) sets aside the issue of global scepticism. It should be noted however that I am raising the spectre of scepticism only to make vivid a point that I take to be obvious, namely, that what ontological options are taken as live (and hence whether the ontological options are rich or poor) depends on certain assumptions that are being made and on the relevant context. Other, less sceptical, stories could bring this point home.

## 6 Cartwright's realist toil: From entities to capacities

1. For the unfamiliar reader, here is a brief statement of some major views. On Lewis's reading, *c* causally explains *e* if *c* is connected to *e* with a network of causal chains. For him, causal explanation consists in presenting portions of explanatory information captured by the causal network. On Woodward's reading, *c* causally explains *e* if *c* and *e* are connected by a relevant (interventionist) counterfactual of the form if *c* hadn't happened, then *e* wouldn't have happened either. On Salmon's reading, *c* causally explains *e* if *c* is connected with *e* by a suitable continuous causal (i.e., capable of transmitting a mark) process. On the standard deductive-nomological reading of causal explanation, for *c* to causally explain *e*, *c* must be a nomologically sufficient condition for *e*. And for Mackie, for *c* to causally explain *e* there must be event-types *C* and *E* such that *C* is an inus-condition for *E*. For details on all these, see Psillos (2002).
2. To see what these worries might be, consider the difference between modest and ambitious transcendental arguments. Is Cartwright's intention to arrive at the modest conclusion *that it is rational to believe* that there is local knowledge, or at the much more ambitious conclusion that *there is* local knowledge?
3. Spurrett (2001) defends a similar point in much more detail.
4. A huge issue here concerns the nature of laws. I favour the Mill–Ramsey–Lewis approach, which I defend in some detail in Psillos (2002, 148–54 & 210–11). This approach can identify laws independently of their ability to support counterfactuals. However, it seems to require some prior notion of

'natural property'. But this notion need not equate properties with causal powers or capacities.

5. For an important survey of the debate around *ceteris paribus* laws, as well as a defence of strict laws in physics, see Earman & Roberts (1999). The interested reader should also see the special issue of *Erkenntnis* (2002, Vol. 57, no 3) on the status of *ceteris paribus* laws.
6. This point is also made vividly in Cartwright's (2002).
7. One might argue that there are clear cases in which a single case is enough to posit a capacity. An example put to me by Christoph Schmidt-Petri is the capacity to run fast: one case is supposed to prove its existence. I am not so sure this is true. What if I run fast (just once) because I took a certain steroid on a given occasion? Surely, in this case I don't have the capacity to run fast, though the steroid might have the (stable) capacity to make people run fast. But this latter capacity would need regular manifestation in order to be posited. For more criticism of Cartwright's argument that capacities are necessary in the methodology of science, see Giere (2008).
8. A variant of this problem has been posed by Morrison (1995).
9. This point calls into question Cartwright's (1999, 90; 2002, 436) claim that capacities show how we can make sense of inductions from single experiments. Undoubtedly, *if* stable capacities are in operation, then knowing them is enough to generalise from a single experiment. But how is the antecedent of the conditional grounded? It seems that we need regular behaviour (and hence plenty of inductive evidence) in order to posit stable capacities in the first place.
10. Compare: something could be an aspirin without having the causal power to produce a white image; but something could *not* be an aspirin without having the power to relieve headaches.
11. See the criticisms of Fisk's views by Aune (1970) and McMullin (1970).
12. A similar point is made by Menzies (2002). Teller (2002, 720–1) also notes that capacities might well be no different from the OK properties that Cartwright argues should figure in laws.
13. This is just one option, of course. As Teller has stressed to me (see also Teller 2002, 722), another option would be to look for a *mechanism* that connects the nature $\varphi$ with its power to produce a characteristic effect in certain circumstances. I have a number of objections to mechanisms that I cannot repeat here; but see Psillos 2004. At any rate, it seems enough for the purposes of this chapter that it remains an *open option* that Humean regularities may get the capacities do whatever they do.

## 7 Is structural realism possible?

1. Hochberg (1994) claims that (a) Russell came to defend a version of 'hypothetico-scientific realism' and (b) an intensional understanding of relations is enough to dispel Newman's challenge. I think that even if (a) is correct, hypothetico-scientific realism is different from structural realism and that an appeal to intensions may be enough to answer the Newman challenge only at the price of abandoning pure structuralism (cf. Psillos 1999, 67–9).
2. For some further discussion see Psillos (2006b).

## 8 *The* structure, the *whole* structure and nothing *but* the structure?

1. French (2006, 174) has claimed that 'the comparison with mathematical structuralism is misleading'. Here is how he thinks an OS should conceive of the matter: 'The quantum structure, say, does not exist independently of any exemplifying concrete system, as in the *ante rem* case, it *is* the concrete system! But that is not to say that such a structure is simply *in re*, because the ontic structural realist does not – or at least should not – accept that the system, composed of objects and relations is prior to the structure. Indeed, the central claim of ontic structure realism is that it is the structure that is both (ultimately) ontically prior and concrete'. I must admit that I find this kind of claim *very* puzzling. To avoid vacuity, where talk about structures is just a roundabout way to talk about actual and concrete physical systems (like a hydrogen molecule), the structuralist should work with a notion of structure that plays two roles. On the one hand, it should be general (or abstract) enough to be independent of concrete physical systems (so that it can be said that it is shared by distinct but structurally similar or identical physical systems; it can be represented mathematically independently of the actual details of concrete physical systems and the like). On the other hand, it should be such that it should be instantiated by (and hence be part of the identity of) concrete physical systems (so that it plays a role in making a physical system what it is; it contributes to the explanation of its causal role and the like). I cannot see how French's claim makes any headway in understanding how these two roles are actually fulfilled by structures as conceived (or should be conceived) by ontic structuralists. He simply assumes what seems convenient for the ontic structuralist, viz., that physical structures (i.e., structures apt to represent physical systems) are concrete and ontically prior (to what, actually?).
2. For more on causation, see Psillos (2002) and (2004).
3. For some discussion of causal structuralism, see Psillos (2006b).
4. French (2006) has tried to answer to most of these objections concerning how causation can be accommodated within ontic structuralism. Though his replies are invariably interesting, his overall strategy seems to fall between two stools. On the one hand, he tends to move towards causal structuralism and on the other, to suggest that the ontic structuralist can mimic the role of objects and their properties in causal relations, by first admitting them *as such* and then by re-conceptualising them as 'mere nodes in a structure' (2006, 181). The first option leaves unaccounted for the causal role attributed to relations. The second option leaves unaccounted for how causal connections emerge from the relations among 'nodes' in a structure. Unless of course, the structure is already causal, which means that the relations among the nodes are already causal. But then, going for the second option, requires a solution to the problem faced by the first option. The nearest French comes to a 'solution' is when he says that 'particular relations, or kinds of relations, [have], as features, causal aspects particular to those relations or kinds' (2006, 182). But this move is simply question-begging. These causal aspects of relations should either be determined by (or supervene upon) the causal aspects of properties and their possessors or a story should be told as to how they emerge and how they are what they are.

## 9 Ramsey's *Ramsey-sentences*

1. The name is inspired from the title of Rae Langton's book *Kantian Humility*. Langton's Kant was epistemically humble because he thought that the intrinsic properties of things-in-themselves are unknowable. I am not claiming that Ramsey was humble in the same way. After I presented the paper on which this chapter is based in Vienna in 2003, D.H. Mellor told me that there was an unpublished paper by the late David Lewis with the title 'Ramseyan Humility'. Stephanie Lewis has kindly provided me with a copy of it. Langton's book is obviously the common source for Lewis's and my *Ramseyan Humility*. But Lewis's *Ramseyan Humility* is different, stronger and more interesting, than mine.
2. Still the best overall account of Ramsey's philosophy of science is Sahlin's (1990, Chapter 5).
3. Here, I disagree with Sahlin's view that Ramsey was an instrumentalist.
4. As he says, in a slightly different context, 'though I *cannot* name particular things of such kinds I can think of there being such things' (1991, 193).
5. Compare what Ramsey says of the blind man who is about to see: 'part of his future thinking lies in his present secondary system' (1931, 261).
6. One of Carnap's lasting, if neglected, contributions in this area is his use of Hilbert's ε-operator as a means to restore some form of semantic atomism compatible with Ramsey-sentences. See Psillos (2000c) for details.
7. This is an inductively established assumption, as Russell took pains to explain (cf. 1927, Chapter 20).
8. Russell agonises a lot about this. He knows that the relation between percepts and stimuli is one-many and not one–one. See (1927, 255–6).
9. See also Mark Sainsbury's excellent (1979, 200–11).
10. More formally, we need a theorem from second-order logic: that every set *A* determines a *full* structure, that is, one which contains all subsets of *A*, and hence every relation-in-extension on *A*. For an elegant and informative presentation of all the relevant proofs, see Ketland (2004).
11. This is not, however, *generally* true. For every theory has a Ramsey-sentence and there are cases of theories whose Ramsey-sentence does not give the isomorphism-class of the models that satisfy the theory. This has been noted by van Benthem (1978, 329 & 324) and has been recently highlighted by Demopoulos (2003b, 395–6).
12. Winnie (1967, 226–7); Demopoulos & Friedman (1985); Demopoulos (2003a, 387); Ketland (2004).
13. In a joint paper (see Appendix IV of Zahar 2001, 243), Zahar and Worrall call the Carnap-sentence 'metaphysical' because it is untestable. What they mean is actually equivalent with what Carnap thought, viz., that the Carnap-sentence has no factual content. They may well disagree with Carnap that it is a meaning postulate. Be that as it may, the Carnap-sentence is part of the content of the original theory TC. So Zahar and Worrall are not entitled to simply excise it from the theory on the grounds that it is metaphysical. The claim that the variables of the Ramsey-sentence range over physical unobservable entities is no less metaphysical and yet it is admitted as part of the *content* of the Ramsey-sentence.
14. This point is defended by Rozeboom (1960).

15. I am not saying that striving for empirical adequacy is a trivial aim. By no means. It is a very demanding – perhaps utopian – aim. What becomes trivial is searching for truth over and above empirical adequacy, since the former is rendered necessary, if the latter holds.
16. For some similar thoughts, see Russell (1927, 216–17).
17. This, however, is what Langton's Kant *denies*. See Langton (1998).
18. This whole issue has been haunted by a claim made by Russell, Schlick, Maxwell and others that intrinsic properties should be directly perceived, intuited, picturable etc. I see no motivation for this, at least any more. Note that this is *not* Lewis's motivation for the thesis that the intrinsic properties of substances are unknowable. For Lewis's reasons see his 'Ramseyan Humility'.
19. For more on this, see Demopoulos (2003a). See also Winnie (1967).
20. Maxwell (1970a,b) as well as Zahar (2001) take Ramsey to have argued that the knowledge of the unobservable is knowledge by description as opposed to knowledge by acquaintance. This, as we have seen, is true. But note that though they go on to argue that this knowledge is purely structural, and that the intrinsic properties of the unobservable are unknowable, this further thesis is independent of the descriptivist claim. So, it requires an *independent* argument. It is perfectly consistent for someone to think that the unobservable is knowable only by means of descriptions and that this knowledge describes its intrinsic properties as well. For an excellent descriptivist account of Ramsey-sentences, see David Papineau (1996).
21. A similar point has been made by Demopoulos (2003b, 398). It is also made by James van Cleve (1999, 157), who has an excellent discussion of how the problem we have discussed appears in Kant, and in particular in an interpretation of Kant's thought as imposing an isomorphism between the structure of the phenomena and the structure of the noumena.
22. In some notes on theories that Ramsey made in August 1929, he seems not to have yet the idea of the theory as an existential judgement. He writes: 'We simply say our primary system can be consistently constructed as part of a wider scheme of the following kind. Here follows dictionary, laws, axioms etc.' (1991, 229).
23. Braithwaite came back to this issue in a critical notice of Eddington's *The Philosophy of Physical Science*. He (1940) argued against Eddington's structuralism based on Newman's point against Russell. He noted characteristically: 'If Newman's conclusive criticism had received proper attention from philosophers, less nonsense would have been written during the last 12 years on the epistemological virtue of pure structure' (1940, 463). Eddington (1941) replied to this point. For a critical discussion of this exchange, see Solomon (1989).
24. Propositional functions can name objects no less than ordinary names, which are normally the subjects of propositions. Hence, ultimately, Ramsey denies *any* substantive distinction between individuals and qualities: 'all we are talking about is two different types of objects, such that two objects, one of each type, could be sole constituents of an atomic fact' (1931, 132). These two types of objects are 'symmetrical' and there is no point in calling one of them *qualities* and the other *individuals*.
25. This might address worries that the Ramsey-sentence involves second-order quantification. For more on this, see Sahlin (1990, 157).

26. I think this is the central message of Lewis's (1984) devastating critique of Putnam's model-theoretic argument against realism.

## 10 Simply the best: A case for abduction

1. These are dealt in detail in Psillos (2002).
2. The relevant literature is really massive. Some important recent items include Kitcher (1981), Lewis (1986) and Salmon (1989).
3. Philosophical attempts to offer circular justifications of ampliative modes of reasoning have been analysed in Psillos (1999, Chapter 4) and in Lipton (2000). See also Chapter 3 of this book.
4. For a rather compelling criticism of the sceptical challenge to induction and of its philosophical presuppositions, see Mellor (1988).
5. It might be claimed that some self-evident beliefs are ampliative and yet certain enough to be the deductive foundations of all knowledge. But (a) it is contentious whether there are such beliefs; and (b) even if there were, they would have to be implausibly rich in content, since deduction cannot create any new content.
6. As Pollock (1986, 39) notes the mere presence of a defeater *R'* is not enough to remove the *prima facie* warrant for a belief Q. For, being itself a reason, *R'* might also be subject to defeaters. Hence, faced with a possible defeater *R'*, we should examine whether *R'* can itself be (or actually is) defeated by other reasons (what Pollock calls 'defeater defeaters').
7. Pollock frames this in terms of the subjunctive conditional: *R* is a reason to deny that *P would be* true unless *Q were* true.
8. This is essentially what Goodman (1954) observed in his notorious 'new riddle of induction'.
9. This may be a bit too strong, since we know that we can always fault the observation. We may, for instance, insist that the observed swan was not really black. Or we may make it part of the meaning of the term 'swan' that all swans are white. On this last move, a black swan cannot really be a swan. But such manoeuvres, though logically impeccable, do not always have the required epistemic force to save the generalisation from refutation. In any case, in EI we know exactly what sort of manoeuvres we have to block in order to render a generalisation rebutted.
10. Goodman-type stories of the form 'All observed emeralds are green. Therefore, all emeralds are grue' involve a different vocabulary between premises and conclusion only in a trivial way. For predicates such as 'grue' are fully definable in terms of the vocabulary of the premises (plus other antecedently understood vocabulary). So, for instance, 'grue' is defined as: 'green if observed before 2010 or blue thereafter'.
11. For a telling critique of hypothetico-deductivism see Laudan (1995). However, Laudan wrongly assimilates inference to the best explanation to hypothetico-deductivism.
12. It may be objected that EI is equally epistemically permissive since, on any evidence, there will be more than one generalisation which entails it. Yet in order to substantiate this claim for the case of EI, one is bound to produce alternative generalisations which either are non-projectible or restate merely

sceptical doubts (e.g., that all ravens are black when someone observes them).

13. For Peirce's views the interested reader should look at Burks (1946), Fann (1970) and Thagard (1981). See also Psillos (2009)
14. Here I am using the word 'probably' with no specific interpretation of the probability calculus in mind. Its use implies only that the conclusion does not follow from the premises in the way that a deductive argument would have it.
15. Someone might press me to explain how the background knowledge can discriminate among competing hypotheses that, if true, would explain a certain explanandum. I think there is no deep mystery here. In a lot of typical cases where reasoners employ IBE, there is just one 'best explanation' that the relevant background knowledge makes possible. Finding it consists in simply searching within the relevant background knowledge. For more on this issue, and for an interesting scientific example, see Psillos (1999, 217–19).
16. For a fuller discussion see Thagard (1988).
17. Note that here I am using the term 'likelihood' informally and not in the statistical sense of it.
18. Failure to discriminate between the likeliest and the loveliest explanation seems to be the reason why Ben-Menahem (1990, 324) claims that '(t)here is nothing particularly deep about the inference to the best explanation. At least there is nothing particularly deep about it qua type of inference.'
19. I have defended the reliability of IBE in some detail in Psillos (1999, 81–90 & 212–22).
20. Normally, we need to eliminate all but one of them (insofar as they are mutually incompatible, of course), but we should surely allow for ties.
21. Here I am leaving aside van Fraassen's (1989) claim that the reasons for acceptance are merely pragmatic rather than epistemic. For a critical discussion of his views see Psillos (1999, 171–6; 2007b) and Harman (1999, Chapter 4).
22. A hypothesis might explain an event without entailing it. It might make it occurrence probable; or it might be such that it makes the occurrence of the event more probable than it was before the explanatory hypothesis was taken into account. More generally, IBE should be able to take the form of statistical explanation either in the form of the Hempelian Inductive-Statistical model (cf. Hempel 1956) or in the form of Salmon's (cf. 1984) Statistical-Relevance model.

## 11 Inference to the best explanation and Bayesianism

1. Niiniluoto (personal communication) has rightly pointed out that there are two strands within Bayesianism. One of them (Levi and Hintikka) promotes the idea of inductive acceptance rules, and hence, it advocates ampliative inferences. The other branch (Carnap, Jeffrey and Howson) rejects acceptance rules and considers only changes of probabilities. It is this latter strand of Bayesianism that I take issue with at this point.
2. There is an interesting idea in Niiniluoto's paper (2004, 72) that needs to be noted. One may call 'systematic power' the explanatory and predictive power

of a hypothesis relative to the total *initial* evidence *e*. Then, one may use this systematic power to determine the prior probability prob(H) of the hypothesis H. Prob(H) will be none other than the *posterior probability of H* relative to its ability to explain and predict the initial evidence *e*, that is relative to its systematic power. In this sense, it can be argued that the prior probability of a hypothesis does depend on its explanatory (i.e., its systematic) power.

3. For more on his, see Psillos (2007b).
4. This line is pressed in a fresh and interesting way in Lipton's (2001). Lipton tries to show how a Bayesian and an Explanationist can be friends. In particular, he shows how explanatory considerations can help in the determination of the likelihood of a hypothesis, of its prior probability and of the relevant evidence. I think all this is fine. But it should be seen not as an attempt at peaceful co-existence, but rather as an attempt to render Bayesianism within the Explanationist framework, and hence as an attempt to render Bayesianism an objectivist theory of confirmation. For more details see Psillos (2007a).

# References

Achinstein, P. (2001). *The Book of Evidence*, New York: Oxford University Press.

Achinstein, P. (2002). 'Is There a Valid Experimental Argument for Scientific Realism?' *The Journal of Philosophy*, **99**: 470–95.

Allison, H.E. (2004). *Kant's Transcendental Idealism: An Interpretation and Defense* (Revised and enlarged edition), New Haven: Yale University Press.

Almeder, R. (1992). *Blind Realism*, New Jersey: Rowman & Littlefield.

Aronson, J.L., Harré, R. and Way, E. (1994). *Realism Rescued*, London: Duckworth.

Aune, B. (1970). 'Fisk on Capacities and Natures', *Boston Studies in the Philosophy of Science*, **8**: 83–7.

Azzouni, J. (1997). 'Thick Epistemic Access: Distinguishing the Mathematical from the Empirical', *The Journal of Philosophy*, **94**: 472–84.

Azzouni, J. (2000). *Knowledge and Reference in Empirical Science*, London and New York: Routledge.

Azzouni, J. (2004). 'Theoretical Terms, Observational Terms, and Scientific Realism', *The British Journal for the Philosophy of Science*, **55**: 371–92.

Bar-Hillel, M. (1980). 'The Base-Rate Fallacy in Probability Judgements', *Acta Psychologica*, **44**: 211–33.

Ben-Menahem, Y. (1990). 'The Inference to the Best Explanation', *Erkenntnis*, **33**: 319–44.

Birnbaum, M.H. (1983). 'Base Rates in Bayesian Inference: Signal Detection Analysis of the Cab Problem', *American Journal of Psychology*, **96**: 85–94.

Blackburn, S. (1984). *Spreading the Word*, Oxford: Oxford University Press.

Bogen, J. and Woodward, J. (1988). 'Saving the Phenomena', *The Philosophical Review*, **97**: 303–52.

BonJour, L. (1985). *The Structure of Empirical Knowledge*, Cambridge MA: Harvard University Press.

Boyd, R. (1981). 'Scientific Realism and Naturalistic Epistemology', in P.D. Asquith and T. Nickles (eds) *PSA 1980*, Vol. 2, East Lansing: Philosophy of Science Association, pp. 613–62.

Boyd, R. (1984). 'The Current Status of the Realism Debate', in J. Leplin (ed.) *Scientific Realism*, Berkeley: University of California Press, pp. 41–82.

Braithwaite, R.B. (1929). 'Professor Eddington's Gifford Lectures', *Mind*, **38**: 409–35.

Braithwaite, R.B. (1940). 'Critical Notice: *The Philosophy of Physical Science*', *Mind*, **49**: 455–66.

Brown, J.R. (1994). *Smoke and Mirrors*, London: Routledge.

Burks, A. (1946). 'Peirce's Theory of Abduction', *Philosophy of Science*, **13**: 301–6.

Carnap, R. (1928). *The Logical Structure of the World*, trans. R. George, Berkeley: University of California Press.

Carnap, R. (1939). 'Foundations of Logic and Mathematics', *International Encyclopaedia of Unified Science*, Chicago: The University of Chicago Press.

Carnap, R. (1963). 'Replies and Systematic Expositions', in P. Schilpp (ed.) *The Philosophy of Rudolf Carnap*, La Salle IL: Open Court, pp. 859–1013.

Cartwright, N. (1983). *How the Laws of Physics Lie*, Oxford: Clarendon Press.
Cartwright, N. (1989). *Nature's Capacities and Their Measurement*, Oxford: Clarendon Press.
Cartwright, N. (1999). *The Dappled World*, Cambridge: Cambridge University Press.
Cartwright, N. (2002). 'In Favour of Laws That Are Not *Ceteris Paribus* After All', *Erkenntnis*, **5**: 425–39.
Chakravartty, A. (2003). 'The Structuralist Conception of Objects', *Philosophy of Science*, **70**: 867–78.
Chakravartty, A. (2007). *A Metaphysics for Scientific Realism – Knowing the Unobservable*, Cambridge: Cambridge University Press.
Churchland, P.M. (1979). *Scientific Realism and the Plasticity of Mind*, Cambridge: Cambridge University Press.
Churchland, P.M. (1985). 'The Ontological Status of Observables', in P.M. Churchland and C.A. Hooker (eds) *Images of Science*, Chicago: The University of Chicago Press, pp. 35–47.
Churchland, P.M. and Hooker, C.A. (eds) (1985). *Images of Science*, Chicago: The University of Chicago Press.
Cleland, C. (2002). 'Methodological and Epistemic Differences Between Historical Science and Experimental Science', *Philosophy of Science*, **69**: 474–96.
Cohen, L.J. (1981). 'Can Human Irrationality Be Experimentally Demonstrated?', *Behavioural and Brain Sciences*, **4**: 317–31.
da Costa, N. and French, S. (2003). *Science and Partial Truth*, New York: Oxford University Press.
Day, T. and Kincaid, H. (1994). 'Putting Inference to the Best Explanation in Its Place', *Synthese*, **98**: 271–95.
Demopoulos, W. and Friedman, M. (1985). 'Critical Notice: Bertrand Russell's *The Analysis of Matter*: Its Historical Context and Contemporary Interest', *Philosophy of Science*, **52**: 621–39.
Demopoulos, W. (2003a). 'On the Rational Reconstruction of Our Theoretical Knowledge', *The British Journal for the Philosophy of Science*, **54**: 371–403.
Demopoulos, W. (2003b). 'Russell's Structuralism and the Absolute Description of the World', in N. Grifin (ed.) *The Cambridge Companion to Bertrand Russell*, Cambridge: Cambridge University Press, pp. 392–418.
Devitt, M. (1984). *Realism and Truth*, Oxford: Blackwell.
Devitt, M. (1997). *Realism and Truth* (Second edition with a new afterword by the author), Princeton: Princeton University Press.
Devitt, M. (2001). 'The Metaphysics of Truth', in M. Lynch (ed.) *The Nature of Truth*, Cambridge MA: MIT Press, pp. 579–612.
Dewey, John (1939). 'Experience, Knowledge and Value: A Rejoinder', in Paul A. Schilpp and Lewis E. Hahn (eds) *The Philosophy of John Dewey*, La Salle: Open Court, pp. 517–608.
Dorling, J. (1992). 'Bayesian Conditionalisation Resolves Positivist/Realist Disputes', *The Journal of Philosophy*, **89**: 362–82.
Douven, I. (1999). 'Inference to the Best Explanation Made Coherent', *Philosophy of Science*, **66**: S424–35.
Dowe, P. (2000). *Physical Causation*, Cambridge: Cambridge University Press.

Dummett, M. (1982). 'Realism', *Synthese*, **52**: 55–112.
Dummett, M. (1991). *Frege: Philosophy of Mathematics*, London: Duckworth.
Earman, J. and Roberts, J. (1999). '*Ceteris Paribus*, There Is no Problem of Provisos', *Synthese*, **118**: 439–78.
Eddington, A.S. (1941). 'Group Structure in Physical Science', *Mind*, **50**: 268–79.
Ellis, B. (1985). 'What Science Aims to Do', in P.M. Churchland and C.A. Hooker (eds) *Images of Science*, Chicago: The University of Chicago Press, pp. 48–74.
Ellis, B. (2005). 'Physical Realism', *Ratio*, **28**: 371–84.
English, J. (1973). 'Underdetermination: Craig and Ramsey', *The Journal of Philosophy*, **70**: 453–62.
Fann, K.T. (1970). *Peirce's Theory of Abduction*, The Hague: Martinus Nijhoff.
Feyerabend, P. (1969). 'Science Without Experience', *The Journal of Philosophy*, **67**: 791–4.
Field, H. (1992). 'Critical Notice: Paul Horwich's *Truth*', *Philosophy of Science*, **59**: 321–33.
Fine, A. (1986a). *The Shaky Game*, Chicago: The University of Chicago Press.
Fine, A. (1986b). 'Unnatural Attitudes: Realist and Instrumentalist Attachments to Science', *Mind*, **95**: 149–79.
Fisk, M. (1970). 'Capacities and Natures', *Boston Studies in the Philosophy of Science*, **8**: 49–62.
Fodor, J. (1991). 'The Dogma that Didn't Bark (A Fragment of a Naturalised Epistemology)', *Mind*, **100**: 201–20.
Forrest, P. (1994). 'Why Most of Us Should Be Scientific Realists: A Reply to van Fraassen', *The Monist*, **77**: 47–70.
Frank, P. (1932). *The Law of Causality and Its Limits* (M. Neurath and R.S. Cohen trans.) Dordrecht: Kluwer.
French, S. (1999). 'Models and Mathematics in Physics', in J. Butterfield and C. Pagonis (eds) *From Physics to Philosophy*, Cambridge: Cambridge University Press, pp. 187–207.
French, S. (2006). 'Structure as a Weapon of the Realist', *Proceedings of the Aristotelian Society*, **106**: 167–85.
French, S. and Ladyman, J. (2003a). 'Remodelling Structural Realism: Quantum Physics and the Metaphysics of Structure', *Synthese*, **136**: 31–65.
French, S. and Ladyman, J. (2003b). 'The Dissolution of Objects: Between Platonism and Phenomenalism', *Synthese*, **136**: 73–7.
Ghins, M. (2001). 'Putnam's No-Miracle Argument: A Critique', in S.P. Clarke and T.D. Lyons (eds) *Recent Themes in the Philosophy of Science*, Dordrecht: Kluwer Academic Publishers, pp. 121–38.
Giere, R. (1988). *Explaining Science: A Cognitive Approach*, Chicago: The University of Chicago Press.
Giere, R. (1999). *Science Without Laws*, Chicago: The University of Chicago Press.
Giere, R. (2008). 'Models, Metaphysics and Methodology', in S. Hartman, C. Hoefer and L. Bovens (eds) *Cartwright's Philosophy of Science*, New York: Routledge, pp. 123–33.
Glennan, S. (1997). 'Capacities, Universality, and Singularity', *Philosophy of Science*, **64**: 605–26.
Glymour, C. (1980). *Theory and Evidence*, Princeton: Princeton University Press.

Goodman, N. (1954). *Fact, Fiction and Forecast*, Cambridge MA: Harvard University Press.

Gower, B. (1998). *Scientific Method: An Historical and Philosophical Introduction*, London: Routledge.

Hanson, N.R. (1958). *Patterns of Discovery: An Inquiry into the Conceptual Foundations of Science*, Cambridge: Cambridge University Press.

Harman, G. (1965). 'Inference to the Best Explanation', *The Philosophical Review*, **74**: 88–95.

Harman, G. (1979). 'Reasoning and Explanatory Coherence', *American Philosophical Quarterly*, **17**: 151–7.

Harman, G. (1986). *Change in View: Principles of Reasoning*, Cambridge MA: MIT Press.

Harman, G. (1995). 'Rationality', in E.E. Smith and D.N. Osherson (eds) *An Invitation to Cognitive Science* Vol. 3 (Thinking), Cambridge MA: MIT Press, pp. 175–211.

Harman, G. (1996). 'Pragmatism and the Reasons for Belief', in C.B. Kulp (ed.) *Realism/Anti-realism and Epistemology*, New Jersey: Rowan & Littlefield, pp. 123–47.

Harman, G. (1999). *Reasoning, Meaning and Mind*, Oxford: Oxford University Press.

Harré, R. (1970). 'Powers', *The British Journal for the Philosophy of Science*, **21**: 81–101.

Harré, R. (1988). 'Realism and Ontology', *Philosophia Naturalis*, **25**: 386–98.

Hawthorne, J. (2001). 'Causal Structuralism', *Philosophical Perspectives*, **15**: 361–78.

Hempel, C. (1958). 'The Theoretician's Dilemma: A Study in the Logic of Theory Construction', *Minnesota Studies in the Philosophy of Science*, **2**, Minneapolis: University of Minnesota Press, pp. 37–98.

Hempel, C. (1965). *Aspects of Scientific Explanation*, New York: The Free Press.

Hempel, C. (1988). 'Provisos: A Problem Concerning the Inferential Function of Scientific Theories', *Erkenntnis*, **28**: 147–64.

Hendry, R. (1995). 'Realism and Progress: Why Scientists Should Be Realists', in R. Fellows (ed.) Philosophy and Technology. Cambridge: Cambridge University Press, pp. 53–72.

Hitchcock, C.R. (1992). 'Causal Explanation and Scientific Realism', *Erkenntnis*, **37**: 151–78.

Hochberg, H. (1994). 'Causal Connections, Universals and Russell's Hypothetico-Scientific Realism', *Monist*, **77**: 71–92.

Horwich, P. (1990). *Truth*, Oxford: Blackwell.

Horwich, P. (1997). 'Realism and Truth', *Poznan Studies in the Philosophy of the Sciences and the Humanities*, **55**, Amsterdam: Rodopi, pp. 29–39.

Howson, C. (2000). *Hume's Problem*, New York: Oxford University Press.

Husserl, E. (1970). *The Crisis of the European Sciences and Transcendental Phenomenology*, Evanston: Northwestern University Press.

Hylton, P. (1994). 'Quine's Naturalism', *Midwest Studies in Philosophy*, **19**: 261–82.

Jardine, N. (1986). *The Fortunes of Inquiry*, Oxford: Clarendon Press.

Jennings, R. (1989). 'Scientific Quasi-Realism', *Mind*, **98**: 225–45.

Josephson, J. *et al* (1994). *Abductive Inference*, Cambridge: Cambridge University Press.

Josephson, J. (2000). 'Smart Inductive Generalisations Are Abductions', in P. Flach and A. Kakas (eds) *Abduction and Induction: Essays on Their Relation and Integration*, Dordrecht: Kluwer Academic Publishers, pp. 31–44.

Jubien, M. (1972). 'The Intensionality of Ontological Commitment', *Nous*, **6**: 378–87.

Ketland, J. (2004). 'Empirical Adequacy and Ramsification', *The British Journal for the Philosophy of Science*, **55**: 287–300.

Kitcher, P. (1981). 'Explanatory Unification', *Philosophy of Science*, **48**: 251–81.

Kitcher, P. (1993a). *The Advancement of Science*, Oxford: Oxford University Press.

Kitcher, P. (1993b). 'Knowledge, Society and History', *Canadian Philosophical Quarterly*, **23**: 155–78.

Koehler, J.J. (1996). 'The Base Rate Fallacy Reconsidered: Descriptive, Normative and Methodological Challenges', *Behavioural and Brain Sciences*, **19**: 1–17.

Korb, K. (2002). 'Bayesian Informal Logic and Fallacy', Melbourne: School of Computer Science and Software Engineering, Monash University, Technical Report 2002/120.

Kukla, A. (1998). *Studies in Scientific Realism*, Oxford: Oxford University Press.

Ladyman, J. (1998). 'What is Structural Realism?' *Studies in History and Philosophy of Science*, **29**: 409–24.

Ladyman, J. (1999). 'Review of Leplin's (1997)', *The British Journal for the Philosophy of Science*, **50**: 181–8.

Ladyman, J. (2001). 'Science, Metaphysics and Structural Realism', *Philosophica*, **67**: 57–76.

Langton, R. (1998). *Kantian Humility*, Oxford: Clarendon Press.

Laudan, L. (1984). *Science and Values*, Berkeley: University of California Press.

Laudan, L. and Leplin, J. (1991). 'Empirical Equivalence and Underdetermination', *The Journal of Philosophy*, **88**: 449–72.

Laudan, L. (1995). 'Damn the Consequences', *The Proceedings and Addresses of the American Philosophical Association*, **6**: 27–34.

Laudan, L. (1996). *Beyond Positivism and Relativism*, Boulder: Westview Press.

Leplin, J. (1997). *A Novel Defence of Scientific Realism*, Oxford: Oxford University Press.

Levin, M. (1984). 'What Kind of Explanation Is Truth?', in J. Leplin (ed.) *Scientific Realism*. Berkeley and Los Angeles: University of California Press, pp. 124–39.

Lewis, D. (1984). 'Putnam's Paradox', *Australasian Journal of Philosophy*, **62**: 221–36.

Lewis, D. (1986). 'Causal Explanation', in his *Philosophical Papers*, Vol. 2, Oxford: Oxford University Press, pp. 214–40.

Lipton, P. (1991). *Inference to the Best Explanation*, London: Routledge.

Lipton, P. (2000). 'Tracking Track Records', *Proceedings of the Aristotelian Society* Suppl. Vol. **74**: 179–205.

Lipton, P. (2001). 'Is Explanation a Guide to Inference?', in G. Hon and S.S. Rakover (eds) *Explanation: Theoretical Approaches and Applications*, Dordrecht: Kluwer, pp. 93–120.

Lycan, W. (1988). *Judgement and Justification*, Cambridge: Cambridge University Press.

Lycan, W. (1989). 'Explanationism, ECHO, and the Connectionist Paradigm', *Behavioural and Brain Sciences*, **12**: 480.

Maxwell, G. (1962). 'The Ontological Status of Theoretical Entities', *Minnesota Studies in the Philosophy of Science*, **3**, Minneapolis: University of Minnesota Press, pp. 3–27.

Maxwell, G. (1970a). 'Theories, Perception and Structural Realism', in R. Colodny (ed.) *The Nature and Function of Scientific Theories*, Pittsburgh: University of Pittsburgh Press, pp. 3–34.

Maxwell, G. (1970b). 'Structural Realism and the Meaning of Theoretical Terms', *Minnesota Studies in the Philosophy of Science*, **4**, Minneapolis: University of Minnesota Press, pp. 181–92.

McMullin, E. (1970). 'Capacities and Natures: An Exercise in Ontology', *Boston Studies in the Philosophy of Science*, **8**: 63–83.

McMullin, E. (1987). 'Explanatory Success and the Truth of Theory', in N. Rescher (ed.) *Scientific Inquiry in Philosophical Perspective*, Lanham: University Press of America, pp. 51–73.

McMullin, E. (1994). 'Enlarging the Known World', in J. Hilgevoord (ed.) *Physics and Our View of the World*, Cambridge: Cambridge University Press, pp. 79–113.

Mellor, D.H. (1988). *The Warrant of Induction*, Cambridge: Cambridge University Press.

Menzies, P. (2002). 'Capacities, Natures and Pluralism: A New Metaphysics for Science?', *Philosophical Books*, **43**: 261–70.

Miller, D. (1974). 'Popper's Qualitative Theory of Verisimilitude', *The British Journal for the Philosophy of Science*, **25**: 166–77.

Miller, R. (1987). *Fact and Method*, Princeton: Princeton University Press.

Monton, B. (ed.) (2007). *Images of Empiricism*. Oxford: Oxford University Press.

Morrison, M. (1995). 'Capacities, Tendencies and the Problem of Singular Causes', *Philosophy and Phenomenological Research*, **55**: 163–8.

Musgrave, A. (1980). 'Wittgensteinian Instrumentalism', *Theoria*, **46**: 65–105.

Musgrave, A. (1988). 'The Ultimate Argument for Scientific Realism', in R. Nola (ed.) *Relativism and Realism in Science*, Dordrecht/Boston: Kluwer, pp. 229–52.

Musgrave, A. (1989). 'NOA's Ark – Fine for Realism', *The Philosophical Quarterly*, **39**: 383–98.

Musgrave, A. (1996). 'Realism, Truth and Objectivity', in R.S. Cohen *et al.* (eds) *Realism and Anti-Realism in the Philosophy of Science*, Dordrecht: Kluwer, pp. 19–44.

Musgrave, A. (1999a). *Essays on Realism and Rationalism*, Amsterdam: Rodopi.

Musgrave, A. (1999b). 'How to Do Without Inductive Logic', *Science & Education*, **8**: 395–412.

Newman, M.H.A. (1928). 'Mr. Russell's 'Causal Theory of Perception'', *Mind*, **37**: 137–48.

Newton-Smith, W.H. (1978). 'The Underdetermination of Theory by Data', *Proceedings of the Aristotelian Society*, Suppl. Vol. **52**: 71–91.

Newton-Smith, W.H. (1987). 'Realism and Inference to the Best Explanation', *Fundamenta Scientiae*, **7**: 305–16.

Newton-Smith, W.H. (1989a). 'Modest Realism' in A. Fine and J. Leplin (eds) *PSA 1988*, Vol. 2, East Lansing: Philosophy of Science Association, pp. 179–89.
Newton-Smith, W.H. (1989b). 'The Truth in Realism', *Dialectica*, **43**: 31–45.
Niiniluoto, I. (1999). *Critical Scientific Realism*, Oxford: Clarendon Press.
Niiniluoto, I. (2004). 'Truth-Seeking by Abduction', in F. Stadler (ed.) *Induction and Deduction in the Sciences*, Dordrecht: Kluwer, pp. 57–82.
Norris, C. (2004). *Philosophy of Language and the Challenge to Scientific Realism*, London: Routledge.
Norton, J. (1994). 'Science and Certainty', *Synthese*, **99**: 3–22.
Papineau, D. (1996). 'Theory-Dependent Terms', *Philosophy of Science*, **63**: 1–20.
Popper, K. (1982). *Realism and the Aim of Science*, London: Hutchinson.
Pollock, J. (1986). *Contemporary Theories of Knowledge*, New Jersey: Rowan & Littlefield.
Pollock, J. (1987). 'Defeasible Reasoning', *Cognitive Science*, **11**: 481–518.
Prior, E., Pargeter, R. and Jackson, F. (1982). 'Three Theses About Dispositions', *American Philosophical Quarterly*, **19**: 251–6.
Psillos, S. (1999). *Scientific Realism: How Science Tracks Truth*, London: Routledge.
Psillos, S. (2000a). 'Abduction: Between Conceptual Richness and Computational Complexity', in A.K. Kakas and P. Flach (eds) *Abduction and Induction: Essays in Their Relation and Integration*, Dordrecht: Kluwer, pp. 59–74.
Psillos, S. (2000b). 'Carnap, the Ramsey-Sentence and Realistic Empiricism', *Erkenntnis*, **52**: 253–79.
Psillos, S. (2000c). 'An Introduction to Carnap's 'Theoretical Concepts in Science'' (together with Carnap's: 'Theoretical Concepts in Science'), *Studies in History and Philosophy of Science*, **31**: 151–72.
Psillos, S. (2001). 'Author's Response', *Metascience* **10**: 366–70.
Psillos, S. (2002). *Causation and Explanation*, Montreal: McGill-Queens University Press.
Psillos, S. (2004). 'A Glimpse of the *Secret Connexion*: Harmonising Mechanisms with Counterfactuals', *Perspectives on Science*, **12**: 288–319.
Psillos, S. (2006a). 'What Do Powers Do When They Are not Manifested?', *Philosophy and Phenomenological Research*, **72**: 135–56.
Psillos, S. (2006b). 'Critical Notice: *Laws in Nature*', *Metascience*, **15**: 454–62.
Psillos, S. (2007a). 'The Fine Structure of Inference to the Best Explanation', *Philosophy and Phenomenological Research*, **74**: 441–8.
Psillos, S. (2007b). 'Putting a Bridle on Irrationality: An Appraisal of van Fraassen's New Epistemology', in B. Monton (ed.) *Images of Empiricism*, Oxford University Press, pp. 134–64.
Psillos, S. (2009). 'An Explorer upon Untrodden Ground: Pierce on Abduction', in J. Woods, D. Gabbay and S. Hartmam (eds) *Handbook of the History of Logic*, Vol. 10, Inductive Logic, Elsevier.
Putnam, H. (1975). *Mathematics, Matter and Method*, Cambridge: Cambridge University Press.
Putnam, H. (1981). *Reason, Truth and History*, Cambridge: Cambridge University Press.
Putnam, H. (1990). *Realism with a Human Face*, Cambridge MA: Harvard University Press.

Quine, W.V. (1955). 'Posits and Reality', in *The Ways of Paradox and Other Essays*, (Revised and enlarged edition 1976), Cambridge, MA: Harvard University Press, pp. 246–54.

Quine, W.V. (1960). *Word and Object*, Cambridge MA: MIT Press.

Quine, W.V. (1975). 'On Empirically Equivalent Systems of the World', *Erkenntnis*, **9**: 313–28.

Quine, W.V. (1981). *Theories and Things*, Cambridge MA: Harvard University Press.

Ramsey, F. (1931). *The Foundations of Mathematics and Other Essays*, R.B. Braithwaite (ed.), London: Routledge and Kegan Paul.

Ramsey, F. (1991). *Notes on Philosophy, Probability and Mathematics*, M.C. Galavotti (ed.), Rome: Bibliopolis.

Reck, E.H. and Price, M.P. (2000). 'Structures and Structuralism in Contemporary Philosophy of Mathematics', *Synthese*, **125**: 341–83.

Rorty, R. (1991). 'Is Science a Natural Kind?', in *Philosophical Papers* Vol. 1, Cambridge: Cambridge University Press.

Rosen, G. (1994). 'What Is Constructive Empiricism?', *Philosophical Studies*, **74**: 143–78.

Rozeboom, W.W. (1960). 'Studies in the Empiricist Theory of Scientific Meaning', *Philosophy of Science*, **27**: 359–73.

Russell, B. (1919). *Introduction to Mathematical Philosophy*. London: George Allen & Unwin.

Russell, B. (1927). *The Analysis of Matter*, London: Routledge and Kegan Paul.

Russell, B. (1948). *Human Knowledge: Its Scope and Limits*, London: Routledge.

Sahlin, N.E. (1990). *The Philosophy of F P Ramsey*, Cambridge: Cambridge University Press.

Sainsbury, M.R. (1979). *Russell*, London: Routledge and Kegan Paul.

Salmon, W. (1984). *Scientific Explanation and the Causal Structure of the World*, Princeton: Princeton University Press.

Salmon, W. (1985). 'Empiricism: The Key Question', in N. Rescher (ed.) *The Heritage of Logical Positivism*, Lanham: University Press of America, pp. 1–21.

Salmon, W. (1989). *Four Decades of Scientific Explanation*, Minneapolis: Minnesota University Press.

Salmon, W. (1997). *Causality and Explanation*, Oxford: Oxford University Press.

Sarkar, H. (1998). 'Review of Leplin's [1997]', *The Journal of Philosophy*, **95**: 204–9.

Schlick, M. (1918/1925). *General Theory of Knowledge* (2nd German edition, A.E. Blumberg trans.), Wien & New York: Springer-Verlag.

Schlick, M. (1932). 'Form and Content: An Introduction to Philosophical Thinking', in Moritz Schlick's *Philosophical Papers*, Vol. II, Dordrecht: Reidel, 1979, pp. 285–369.

Sellars, W. (1956). 'Empiricism and the Philosophy of Mind', in *Minnesota Studies in the Philosophy of Science*, **1**, Minneapolis: University of Minnesota Press, pp. 253–329.

Sellars, W. (1963). *Science, Perception and Reality*, Atascadero CA: Ridgeview Publishing Company.

Sellars, W. (1977). *Philosophical Perspectives: Metaphysics and Epistemology*, Atascadero CA: Ridgeview Publishing Company.

Shapiro, S. (1997). *Philosophy of Mathematics: Structure and Ontology*, Oxford: Oxford University Press.
Shoemaker, S. (1980). 'Causality and Properties', in Peter van Inwagen (ed.) *Time and Change*, Reidel, pp. 109–35.
Smart, J.J.C. (1963). *Philosophy and Scientific Realism*, London: Routledge and Kegan Paul.
Smith, P. (1998). 'Approximate Truth and Dynamical Theories', *The British Journal for the Philosophy of Science*, **49**: 253–77.
Sober, E. (2002). 'Bayesianism – Its Scope and Limits', *Proceedings of the British Academy*, **113**: 21–38.
Solomon, G. (1989). 'An Addendum to Demopoulos and Friedman (1985)', *Philosophy of Science*, **56**: 497–501.
Spurrett, D. (2001). 'Cartwright on Laws of Composition', *International Studies in the Philosophy of Science*, **15**: 253–68.
Stanford, P. Kyle (2006). *Exceeding Our Grasp: Science, History, and the Problem of Unconceived Alternatives*, Oxford: Oxford University Press.
Suppe, F. (1989). *The Semantic Conception of Theories and Scientific Realism*, Chicago: University of Illinois Press.
Taylor, B. (1987). 'The Truth in Realism', *Revue Internationale de Philosophie*, **160**: 45–63.
Teller, P. (2002). 'Review of *The Dappled World*', *Nous*, **36**: 699–725.
Thagard, P. (1978). 'Best Explanation: Criteria for Theory Choice', *The Journal of Philosophy*, **75**: 76–92.
Thagard, P. (1981). 'Peirce on Hypothesis and Abduction', in K. Ketner *et al.* (ed.) *C. S. Peirce Bicentennial International Congress*, Texas: Texas Tech University Press, pp. 271–4.
Thagard, P. (1988). *Computational Philosophy of Science*, Cambridge MA: MIT Press.
Thagard, P. (1989). 'Explanatory Coherence', *Behavioural and Brain Sciences*, **12**: 435–502.
Tichy, P. (1974). 'On Popper's Definition of Verisimilitude', *The British Journal for the Philosophy of Science*, **25**: 155–60.
Trout, J.D. (1998). *Measuring the Intentional World*, Oxford: Oxford University Press.
Turner, D. (2007). *Making Prehistory: Historical Science and the Scientific Realism Debate*, Cambridge: Cambridge University Press.
Tversky, A. and Kahneman, D. (1982). 'Judgment under Uncertainty: Heuristics and Biases', in D. Kahneman, P. Slovic and A. Tversky (eds) *Judgment Under Uncertainty: Heuristics and Biases*, Cambridge: Cambridge University Press, pp. 3–22.
van Benthem, J.F.A.K. (1978). 'Ramsey Eliminability', *Studia Logica*, **37**: 321–36.
van Cleve, J. (1999). *Themes From Kant*, Oxford: Oxford University Press.
van Fraassen, B.C. (1980). *The Scientific Image*, Oxford: Clarendon Press.
van Fraassen, B.C. (1985). 'Empiricism in Philosophy of Science', in P.M. Churchland and C.A. Hooker (eds) *Images of Science*, Chicago: The University of Chicago Press, pp. 245–308.
van Fraassen, B.C. (1989). *Laws and Symmetry*, Oxford: Clarendon Press.
van Fraassen, B.C. (1994). 'Against Transcendental Empiricism', in T.J. Stapleton (ed.) *The Question of Hermeneutics*, Dordrecht: Kluwer, pp. 309–36

van Fraassen, B.C. (2000a). 'The False Hopes of Traditional Epistemology', *Philosophy and Phenomenological Research*, **60**: 253–80.

van Fraassen, B.C. (2000b). 'Michel Ghins on the Empirical Versus the Theoretical', *Foundations of Physics*, **30**: 1655–61.

van Fraassen (2006). 'Structure: Its Shadow and Substance', *The British Journal for the Philosophy of Science* **57**: 275–307.

Weyl, H. (1963). *Philosophy of Mathematics and Natural Science*, New York: Atheneum.

Winnie, J. (1967). 'The Implicit Definition of Theoretical Terms', *The British Journal for the Philosophy of Science*, **18**: 223–9.

Winnie, J. (1970). 'Theoretical Analyticity', *Boston Studies in the Philosophy of Science*, **8**: 289–305.

Windschitl, P.D, and Wells, G.L. (1996). 'Base Rates Do not Constrain Nonprobability Judgments', *Behavioural and Brain Sciences*, **19**: 40–1.

Worrall, J. (1985). 'Scientific Discovery and Theory-Confirmation', in J. Pitt (ed.) *Change and Progress in Modern Science*, Dordrecht: Reidel, pp. 301–31.

Worrall, J. (1989). 'Structural Realism: The Best of Both Worlds?', *Dialectica*, **43**: 99–124.

Wright, C. (1986). 'Scientific Realism, Observation and the Verification Principle', in G. Macdonald and C. Wright (eds) *Fact, Science and Morality*, Oxford: Blackwell, pp. 247–74.

Wright, C. (1988). 'Realism, Antirealism, Irrealism, Quasi-Realism', *Midwest Studies in Philosophy*, **12**: 25–49.

Wright, C. (1992). *Truth and Objectivity*, Cambridge MA: Harvard University Press.

Zahar, E. (2001). *Poincaré's Philosophy: From Conventionalism to Phenomenology*, La Salle IL: Open Court.

# Index